T0231347

SECOND EDITION # The Food Chemistry Laboratory

A Manual for Experimental Foods, Dietetics, and Food Scientists

CRC Series in CONTEMPORARY FOOD SCIENCE

Fergus M. Clydesdale, Series Editor
University of Massachusetts, Amherst

Published Titles:

*America's Foods Health Messages and Claims:
Scientific, Regulatory, and Legal Issues*
James E. Tillotson

New Food Product Development: From Concept to Marketplace
Gordon W. Fuller

Food Properties Handbook
Shafiur Rahman

*Aseptic Processing and Packaging of Foods:
Food Industry Perspectives*
Jarius David, V. R. Carlson, and Ralph Graves

*The Food Chemistry Laboratory: A Manual for Experimental Foods, Dietetics,
and Food Scientists, Second Edition*
Connie M. Weaver and James R. Daniel

Handbook of Food Spoilage Yeasts
Tibor Deak and Larry R. Beauchat

Food Emulsions: Principles, Practice, and Techniques
David Julian McClements

*Getting the Most Out of Your Consultant: A Guide
to Selection Through Implementation*
Gordon W. Fuller

Antioxidant Status, Diet, Nutrition, and Health
Andreas M. Papas

Food Shelf Life Stability
N.A. Michael Eskin and David S. Robinson

Bread Staling
Pavinee Chinachoti and Yael Vodovotz

Interdisciplinary Food Safety Research
Neal M. Hooker and Elsa A. Murano

*Automation for Food Engineering: Food Quality Quantization
and Process Control*
Yanbo Huang, A. Dale Whittaker, and Ronald E. Lacey

SECOND EDITION

The Food Chemistry Laboratory

A Manual for Experimental Foods, Dietetics, and Food Scientists

Connie M. Weaver
James R. Daniel

Department of Foods and Nutrition
Purdue University
West Lafayette, Indiana

CRC PRESS

Boca Raton   London   New York   Washington, D.C.

Library of Congress Cataloging-in-Publication Data

Weaver, Connie, 1950-
    The food chemistry laboratory : a manual for experimental foods, dietetics, and food
scientists. — 2nd ed. / Connie Weaver and James Daniel.
        p. cm. — (Contemporary food science)
    Includes index.
    ISBN 0-8493-1293-0 (alk. paper)
    1. Food—Analysis—Laboratory manuals. 2. Food—Composition—Laboratory manuals. I.
Daniel, James. II. Title. III. CRC series in contemporary food science.

TX541 .W43 2003
664'.07—dc21   2002038797

Visit the CRC Press Web site at www.crcpress.com

© 2003 by CRC Press LLC

No claim to original U.S. Government works
International Standard Book Number 0-8493-1293-0
Library of Congress Card Number 2002038797
5 6 7 8 9 0

■ FOREWORD

These laboratory exercises have been designed to illustrate some of the chemical and physical principles discussed in lectures. The laboratory experience should provide a more detailed knowledge of methods and equipment used in food research and also should provide an opportunity for the student to become familiar with the main journals in which food research is reported. The student will learn to keep a laboratory notebook; in addition, the student will become familiar with the fundamentals of designing, executing, and reporting the results of a research project.

The student is expected to read the laboratory procedures before class so that he or she may perform experiments more efficiently and understand the reason for the results obtained. A clean uniform or lab coat must be worn in the laboratory. Hair must be controlled when sensory evaluation is involved. All equipment must be cleaned and stored properly after experimentation.

Any laboratory accident must be reported immediately to the instructor. Locate the carbon dioxide fire extinguisher and be sure you know how to use it. Do not take any chances when a fire starts: use the extinguisher.

We are grateful to Karen Jamesen for her contributions to the laboratory on pectin and to Elton Aberle and John Forrest for their contributions to the myoglobin experiment.

C.M. Weaver
West Lafayette, Indiana

J.R. Daniel
West Lafayette, Indiana

THE AUTHORS

Connie Weaver, Ph.D., is professor and head of foods and nutrition at Purdue University, West Lafayette, Indiana. She joined Purdue in 1978 and became head of the department in 1991.

Dr. Weaver grew up in Oregon. All three of her degrees are in food science and human nutrition with minors in chemistry and plant physiology. She received her B.S. and M.S. from Oregon State University and her Ph.D. from Florida State University.

Dr. Weaver's research is on minerals important to human health. Current projects include (1) chemical form of minerals in foods, (2) mineral bioavailability, (3) calcium metabolism in adolescents, (4) exercise and bone mass in young women, and (5) exercise and iron status in young women. Dr. Weaver has contributed more than 150 research publications and book chapters. Dr. Weaver has been the recipient of many research grants from the National Institutes of Health, the U.S. Department of Agriculture, and various commodity groups and industries.

For her contributions in teaching food chemistry, Dr. Weaver was awarded Purdue University's Outstanding Teaching Award and the school's Mary L. Matthews Teaching Award. She has served as a scientific lecturer and on the executive committee for the Institute of Food Technologists. She is past president of the American Society for Nutritional Sciences and is on the board of trustees of the International Life Sciences Institute. She is on the editorial boards of the *American Journal of Clinical Nutrition*, the Academic Press Food Science and Technology Book Series, and *Advances in Food and Nutrition Research*. She is also a member of the American Chemical Society, the American Association for Advancement of Science, the Society for Experimental Biology and Medicine, and the American Society for Bone and Mineral Research. She is a fellow of the American College of Nutrition.

Dr. Weaver and Dr. Daniel are coauthors of the "Functional Carbohydrates" chapter in *Food Chemistry: Principles and Applications*, published by Science Technology Systems in 2000. Dr. Weaver co-authored the third edition of *Foods: A Scientific Approach* with Helen Charley; this book was published by Prentice-Hall in 1998.

James Daniel, Ph.D., is Associate Professor of Foods and Nutrition at Purdue University, West Lafayette, Indiana. He joined Purdue in 1980.

Dr. Daniel grew up in Kansas. His degrees are in chemistry. He received his B.A. from Kansas State Teachers College (now Emporia State University) and his Ph.D. from Texas A&M University.

Dr. Daniel's research is in the area of carbohydrates. Specifically, he has interests in low-calorie sucrose replacers derived from low-molecular-weight carbohydrates, low-calorie fat replacers derived from high-molecular-weight carbohydrates, discovery and use of food gums to control the texture of foods, and Maillard browning in foods. Dr. Daniel has contributed to more than 35 research publications and book chapters. Dr. Daniel is a member of the Institute of Food Technologists and co-authored the "Functional Carbohydrates" chapter (with Dr. Weaver) in *Food Chemistry: Principles and Applications*, published by Science Technology Systems in 2000.

TABLE OF CONTENTS

1 LITERATURE SEARCH

■ ABSTRACTS AND INDEXES

Agriculture Index
Applied Science and Technology Index
Biological Abstracts
Chemical Abstracts
Food Science and Technology Abstracts
Science Citation Index

■ JOURNALS

Advances in Food Research
Agriculture and Biological Chemistry
American Egg and Poultry Review
American Fruit Grower
American Potato Journal
Baker's Digest
Canadian Institute of Food Science and Technology Journal
Carbohydrate Polymers
Carbohydrate Research
Cellulose Chemistry and Technology
Cereal Chemistry
Cereal Foods World
Food Chemistry
Food Engineering
Food Product Development
Food Technology
Home Economics Research Journal
Journal of Agricultural and Food Chemistry
Journal of the Association of Official Analytical Chemists

Journal of Animal Science
Journal of the American Dietetic Association
Journal of Dairy Research (British)
Journal of Food Biochemistry
Journal of Food Engineering
Journal of Food Protection
Journal of Food Science
Journal of Plant Foods
Journal of the Science of Food and Agriculture
Meat Science
Poultry Science
Proceedings of the American Society of Horticultural Science
Quick Frozen Foods
United States Egg and Poultry Magazine
World Poultry Science Journal

■ ADVANCES AND REVIEWS

Advances in Carbohydrate Chemistry
Advances in Colloid Science
Advances in Enzymology
Advances in Food Research
Advances in Lipid Research
Agricultural Institute Review
American Dairy Review
American Egg and Poultry Review
Critical Reviews in Food Science and Nutrition
Nutrition Abstracts and Reviews
Recent Advances in Food Science

■ GENERAL

Amerine, M.A., Pangborn, R.A., and Roessler, E.B., *Principles of Sensory Evaluation of Food,* Academic Press, New York, 1965.

A.O.A.C., *Official Methods of Analysis,* 17th ed., Association of Official Analytical Chemists, Washington, D.C., 2002.

ASTM Committee E-18, *Guidelines for the Selection and Training of Sensory Panel Members,* STP 758, American Society for Testing and Materials, Philadelphia, 1981.

Bennion, M., *The Science of Food,* Harper & Row, San Francisco, 1980.

Bourne, M.C., *Food Texture and Viscosity: Concept and Measurement,* Academic Press, New York, 1982.

Charley, H., *Food Science,* 2nd ed., John Wiley & Sons, New York, 1982.

Charley, H. and Weaver, C.M., *Foods: A Scientific Approach,* Merrill/Prentice Hall, Indianapolis, 1998.

Christen, G. and Smith, J.S., *Food Chemistry: Principles and Applications,* Science Technology Systems, West Sacramento, CA, 2000.

Cohen, S.H., *Studies of Food Microstructure,* Scanning Electron Microscopy, Chicago, 1982.

Critical Reviews in Food Science and Nutrition (journal).

DeMan, J.M., *Principles of Food Chemistry,* 3rd ed., Aspen Publishers, Gaithersburg, MD, 1999.

Dickenson, E., *Colloids in Food,* Applied Science Publishing, New York, 1982.

Feeney, R.E. and Whitaker, J.R., *Modification of Proteins: Food Nutritional and Pharmacological Aspects,* Advances in Chemistry Series No. 198, American Chemical Society, Washington, D.C., 1982.

Fennema, O.R., *Food Chemistry,* 3rd ed., Marcel Dekker, New York, 1996.

Finley, J., *Chemical Changes in Food During Processing,* AVI Publishing, Westport, CT, 1985.

Francis, F. and Clydesdale, F.M., *Food Colorimetry,* AVI Publishing, Westport, CT, 1975.

Furia, T., Ed., *CRC Handbook of Food Additives,* 2nd ed., CRC Press, Boca Raton, FL, 1972.

Glicksman, M., Ed., *Food Hydrocolloids,* Vols. I, II, and III, CRC Press, Boca Raton, FL, 1986.

Glicksman, M., *Gum Technology in the Food Industry,* Academic Press, New York, 1969.

Heimann, W., *Fundamentals of Food Chemistry,* AVI Publishing, Westport, CT, 1980.

Lee, F.A., *Basic Food Chemistry,* AVI Publishing, Westport, CT, 1983.

McWilliams, M., *Foods: Experimental Perspectives,* Macmillan, New York, 1993.

Maerz, A. and Paul, M.R., *Dictionary of Color,* McGraw-Hill, New York, 1950.

Nielsen, S.S., *Food Analysis,* 2nd ed., Aspen Publishers, Gaithersburg, MD, 1999.

Paul, P.C. and Palmer, H.H., *Food Theory and Applications,* John Wiley & Sons, New York, 1972.

Phyllips, G.O., Wedlock, D.J., and Williams, D.A., Eds., *Gums and Stabilizers for the Food Industry: Interactions of Hydrocolloids,* Pergamon Press, Elmsford, NY, 1982.

Potter, N.N., *Food Science,* Routledge, Chapman and Hall, Incorporated, Georgetown, Ontario, 1986.

Szczesniak, A.S., Branst, M.A., and Friedman, H.H., Development of standard rating scales for mechanical parameters of texture and correlation between the objective and the sensory methods of texture evaluation, *J. Food Sci.,* 1963; 28:397–403.

Taylor, R.S., *Food Additives,* John Wiley & Sons, New York, 1980.

Walstra, P. and Jenness, R., *Dairy Chemistry and Physics,* John Wiley & Sons, New York, 1984.

Watts, B.M., Ylimake, G.L., Jeffery, L.E., and Elias, L.G., *Basic Sensory Methods for Food Evaluation,* International Development Research Centre, Ottawa, 1989.

■ INTERNET SOURCES OF INFORMATION

Institute of Food Technologists — http://www.ift.org

World Food Net — http://www.worldfoodnet.com/

Food Resource — http://www.orst.edu/food-resource/food.html

Food safety information — http://www.foodsafety.gov/
Internet resources in foods and nutrition — http://www.dfst.csiro.au/fdnet20a.htm
Shockwave animations of food chemistry processes — www2.hawaii.edu/lynn/main.html
American Dietetic Association — http://www.eatright.org/
Food and Drug Administration — http://www.fda.gov/

2 EVALUATION OF FOODS

Quality of food is often assessed in terms of three main parameters: color, texture, and flavor.

■ COLOR

Visual perception of color results from activation of the retina by electromagnetic waves in the visible spectrum (see Figure 2.1). Two widely used systems for objectively describing color are the Munsell and C.I.E. color systems.

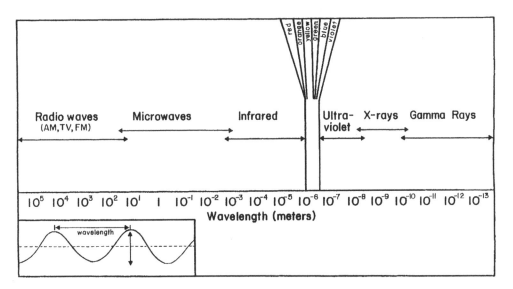

Figure 2.1 Electromagnetic spectrum (λ in nm).

The Munsell color notation is based on a tridimensional color space involving three attributes of color: hue, value, and chroma.

The hue scale is based on five primary and five intermediate hues: red, yellow, green, blue, purple, yellow red, green yellow, blue green, purple blue, and red purple.

The value scale is a lightness scale ranging from black (0) to white (10).

The chroma scale is a measure of the departure of the color perceived from gray of the same lightness (neutral gray = 0).

The color of an object may be matched to a chart of various hues arranged in rows in accordance with the chroma scale and in columns according to value scales (see Munsell, A.H., *A Color Notation*,

Munsell Color Co., Baltimore, 1947). Other types of color dictionaries are also available; the most widely used is the *Dictionary of Color* (Maerz, A. and Paul, M.R., McGraw-Hill, New York, 1960).

The C.I.E. system of color is a more objective means of specifying color in accordance with the recommendations of the International Commission on Illumination in terms of a tristimulus system, a defined standard observer, and a particular coordinate system. In the C.I.E. system, color is specified in terms of the three primary values: amber — X, green — Y, and blue — Z. These values can be obtained with reflectance meters or spectrophotometers. From the tristimulus values, the proportions of each primary are calculated as the following ratios:

$$x = \frac{X}{X + Y + Z} \qquad y = \frac{Y}{X + Y + Z} \qquad z = \frac{Z}{X + Y} = Z$$

These are referred to as the chromaticity coordinates or trichromatic coefficients. Since $x + y + z = 1$, it is sufficient to use $x + y$ to define chromaticity. The x and y coordinates can be plotted on a chromaticity diagram (see Figure 2.2), and the color of the test food can be located in color and space.

A third system, the Hunter color solid (see Figure 2.3), attempts to reconcile differences between the Munsell and C.I.E. color spaces. Color parameters are L, ±a, and ±b, where L is visual lightness, which is similar to value in the Munsell system or Rd (luminous reflectance) in the C.I.E. system. The light reflection is read as + (plus) or − (minus) a and b.

■ TEXTURE

Physical characteristics of foods provide a variety of tactile stimuli that are sometimes difficult to classify; thus, precise terminology is important in preparing score cards for sensory evaluation. One must ask the taste panel to evaluate the specific characteristic(s) of interest to the researcher. The researcher must also be able to identify which aspect of texture is reflected in the various objective techniques. One useful system for classifying physical properties has been proposed by Szczesniak et al., as outlined in Table 2.1.

■ FLAVOR

Flavor characteristics include taste and odor.

Taste sensations are produced as substances dissolved in the saliva interact with the taste buds in the papillae on the tongue. The four primary tastes are sweet, salty, sour, and bitter. A fifth primary taste, umami, has been proposed and is described as a meaty, savory taste.

Odor sensations are perceived when volatile substances contact the olfactory epithelium in the upper part of the nasal cavity. Odor is much more complex to classify than taste due to the seemingly infinite number of subtle differences and to the limited vocabulary available to describe these sensations. Crocker and Henderson attempted to define odors in terms of a four-modular classification: fruity, acid or sharp, blunt or tarry or scorched, and caprylic or goat-like. Each factor is rated on an eight-point scale. Schultz uses a nine-modular system: fragrant, burnt, goaty, etherish, sweet, rancid, oily, metallic, and spicy. Amoore (Amoore, J., Stereochemical and vibrational theories of odor, *Nature* 1971;233:270–271) suggested a seven-module system consisting of ethereal, camphoraceous, musky, floral, minty, pungent, and putrid.

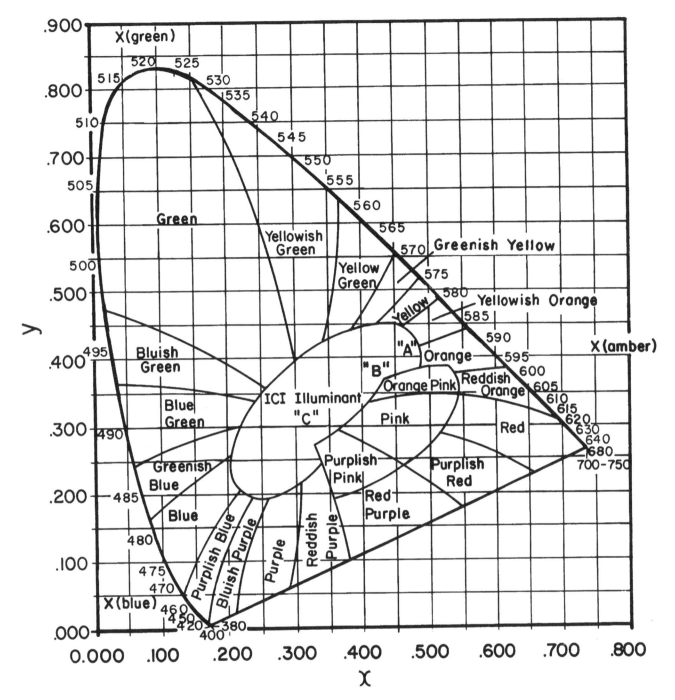

Figure 2.2 C.I.E. chromaticity diagram.

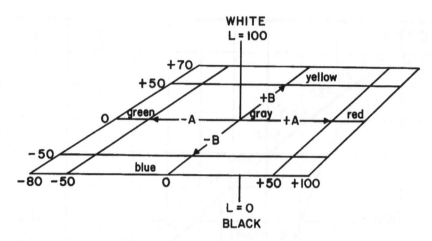

Figure 2.3 Diagrammatic sketch of the Hunter color solid. The vertical direction refers to darkness–lightness (0–100). One horizontal axis refers to +a (redness) and –a (greenness); the other axis refers to +b (yellowness) and –b (blueness).

TABLE 2.1

Relations Between Textural Parameters and Popular Nomenclature

Mechanical Characteristics

Primary Parameters	Secondary Parameters	Popular Terms
Hardness		Soft–firm–hard
Cohesiveness	Brittleness	Crumbly–crunchy–brittle
	Chewiness	Tender–chewy–tough
	Gumminess	Short–mealy–pasty–gummy
Viscosity		Thin–viscous
Elasticity		Plastic–elastic
Adhesiveness		Sticky–tacky–gooey

Geometric Characteristics

Class	Examples
Particle size and shape	Gritty, grainy, coarse, etc.
Particle shape and orientation	Fibrous, cellular, crystalline, etc.

Other Characteristics

Primary Parameters	Secondary Parameters	Popular Terms
Moisture content	Oiliness	Dry–moist–wet–watery
Fat content	Greasiness	Oily
		Greasy

3 OBJECTIVE METHODS

I. Color
   A. Spectrophotometer
   B. Reflection meters or color difference meters
   C. Color system (see Paul and Palmer, 1972; Francis and Clydesdale, 1975)
      1. C.I.E.
      2. Munsell
      3. Hunter color solid
   D. Match to color chips (see Maerz and Paul, *Dictionary of Color*)
   E. Hunter colorimeter

II. Tenderness
   A. Penetrometer
   B. Compressometer
   C. Shortometer
   D. Shear press
   E. Texture analyzer

III. Volume
   A. Seed displacement
   B. Compensating polar planimeter
   C. Standing heights

IV. Juiciness

V. Cell structure
   A. Photography
   B. Ink prints/photocopy
   C. Histological
   D. Preserving sample

VI. Viscosity
   A. Pipette
   B. Jelmeter
   C. Viscometer
   D. Consistometer
   E. Linespread test
   F. Rheometer
   G. VISCO/amylo/GRAPH
   H. RapidViscoAnalyzer (RVA)

   VII. Temperature
        A. Thermometers
        B. Thermistors

  VIII. pH
        A. pH meter

   IX. Water activity
        A. Decagon CX-2 water activity system

    X. Specific gravity
        A. Weight of known volume
        B. Hydrometer
        C. Pycnometer

   XI. Foam stability
        A. Funnel test

  XII. Shrinkage
        A. Change in size
        B. Change in weight

4 SENSORY METHODS

I. Types of sensory tests
   A. Difference tests
      1. Paired comparison: simple difference
      2. Paired comparison: directional difference
      3. Duo-trio
      4. Triangle
   B. Rank order
   C. Rating differences
   D. Descriptive analysis
      1. Category scaling: structured
      2. Category scaling: unstructured
      3. Ratio scaling or magnitude estimation
   E. Threshold
   F. Affective tests
      1. Paired preference
      2. Ranking
      3. Hedonic rating scale

II. Panel
   A. Type
      1. Trained: use 3–10
      2. Semi-trained: use 8–25
      3. Untrained or consumer: use >80
   B. Selection criteria
   C. Composition (age, sex)

III. Material evaluated
   A. Preparation
      1. Method
      2. Carrier (if used)
   B. Presentation
      1. Coding
      2. Order of serving
      3. Sample size
      4. Temperature and method of control
      5. Sample container and utensils used
      6. Time of day
      7. Special conditions (time interval, mouth rinsing, etc.)
      8. Special instructions to panelists

IV. Statistical design
  A. Type of experiment (randomized block, factorial, etc.)
  B. Replications

V. Environmental conditions
  A. Setting (booth, store)
  B. Lighting (color)

VI. Data analysis

■■ DEVELOPING FORMS FOR SENSORY TESTS

Basic Considerations

I. General format
  A. One form for each judge for each set of samples
  B. Space for name, date, set
  C. Directions
  D. Sample code
  E. Space for recording response

II. Development of realistic scorecard
  A. Place factors on score sheet in logical order.
    1. Sight
    2. Odor
    3. Those judged orally
  B. Scale should be easily analyzed statistically when presenting data graphically; the highest number should represent the most desirable.

III. Common difficulties with scoring
  A. Tendency to be too arbitrary — false sense of exactness
  B. Contrast error — psychological tendency when scoring samples of very low and very high quality
  C. Central tendency — psychological error that is frequently observed when extreme values are seldom used
  D. Tendency of subjective scales to drift in meaning with time and with judges — can be avoided by using control or standard

Examples of Sensory Test Forms

CHOCOLATE CAKE

Name_____ Date _____

Group _____ Sample nos._____

Please analyze each group of samples and rate each sample for the different qualities listed. If there is no difference between samples, it is permissible to give them the same rating.

Grain	**Uniformity of Cells**	**Lightness**
____ thin cell walls	____ uniform cells	____ light and fluffy
____	____	____
____	____	____
____	____	____
____ thick cell walls	____ irregular cells	____ compact

Moistness of Crumb	**Flavor**	**Overall Desirability**
____ moist and velvety	____ fresh	____ most desirable
____	____	____
____	____	____
____	____	____
____ dry	____ stale	____ least desirable

This sensory test form is an example of an analytical tool designed to identify specific characteristics of a sample. If the objective is to determine the preferences of a sample parameter rather than to describe the parameter, a test form using a hedonic scale would be a better choice. Similar descriptions could be used for many characteristics including flavor, texture/tenderness, color, juiciness, and overall acceptance. An example using flavor follows:

PRODUCT_____

Panelist_____ Date_____

Judge each sample for the characteristic listed and check how much you like or dislike each. Use the appropriate space to show your attitude by checking at the point that best describes your feeling about the sample.

Characteristic: FLAVOR	SAMPLE NUMBER
Like extremely	
Like very much	
Like moderately	
Like slightly	
Neither like nor dislike	
Dislike slightly	
Dislike moderately	
Dislike very much	
Dislike extremely	
Comments:	

If the researcher wishes to know the darkness of a cake as determined by a sensory evaluation, the first type of form could be used. If, however, the researcher wants to know the preferred sample regardless of how light or dark it is, the second type of form could be used. Additional examples may be found in the laboratory exercises on sensory evaluation.

5

LABORATORY NOTEBOOK

Laboratory results must be placed in a laboratory notebook that does not have easily removable pages. Small slips of paper used for recording data are easily lost. Pages must not be torn from the laboratory notebook. If an error is made, cross it out. The instructor will examine laboratory notebooks periodically to aid the student in recording data properly.

■ FORMAT

I. Title, date, unusual laboratory conditions

II. Purpose

III. Experimental procedures — Do not copy lab manual. Record only changes in procedures or expanded descriptions of procedures.

IV. Results — The experimental data may be presented in tabular or graphical form or both. It saves valuable lab time if tables and graphs are prepared before the laboratory class.
A. Tables (refer to Tables 5.1 and 5.2)
  1. Number each table in sequence.
  2. Each table must have a title.
  3. Draw lines only as shown.
  4. Units of measure must be shown in the heading or elsewhere in the table.
  5. Objective and subjective data should be presented in separate tables.
  6. Use appropriate footnotes.
B. Figures (refer to Figures 5.1–5.3)
  1. Number each figure in sequence.
  2. Each figure must have a title, which should be concise but more than a mere repetition of axes labels.
  3. Use graph paper unless notebook has appropriate squares.
  4. Each axis must be labeled and units of measure given.
  5. Independent variables are placed on the x-axis (abscissa) and dependent variables are placed on the y-axis (ordinate). Generally, use a line graph when the independent variable is continuous, and a bar graph when the independent variable is discontinuous.

V. Discussion — The discussion of the results includes a statement of the results as given in the tables and figures. Do refer to each table and figure in your discussion. Also include the errors involved in experimentation, difficulties encountered, possible explanations for results obtained, and any conclusions that could be drawn from your results. Be sure to draw comparisons between objective and subjective results when possible. When questions are included with experiments in the laboratory manual, incorporate answers to these questions in the discussion. Use references in your explanation, such as "Kim and Wang (2001) studied formation of inulin gels…" or "…a kinetic study of thermally induced inulin gels was reported (Kim and Wang 2001)."

VI. References — Use the style of the *Journal of Food Science*. An example of a journal article is "Kim Y., Wang S.S. 2001. Kinetic study of thermally induced inulin gel. J Food Sci 66:991–997." To facilitate location of a particular laboratory write-up, number the pages of the laboratory notebook consecutively and prepare a table of contents at the beginning of the notebook.

■ SAMPLE TABLES AS REPORTED IN THE *JOURNAL OF FOOD SCIENCE*

TABLE 5.1

Average Composition of 490 Samples of Honey and Range of Values (White and others 1962)

Component	Average	Standard Deviation	Range
Moisture, %	17.2	1.46	13.4–22.9
Fructose, %	38.19	2.07	27.25–44.26
Glucose, %	31.28	3.03	22.03–40.75
Sucrose, %	1.31	3.03	0.25–7.57
Maltose, %	7.31	2.09	2.74–15.98
Higher sugars, %	1.50	1.03	0.13–8.49
Free acid, meq/kg	22.03	8.22	6.75–47.19
Lactone, meq/kg	0.335	0.135	0.0–0.95
Ash, %	0.169	0.15	0.02–1.028
Nitrogen, %	0.041	0.026	0.0–0.133
Diastase value	20.8	9.76	2.1–61.2

Tables that give subjective data for variations are written in the same manner as tables that present objective data. For example, the formats of Table 5.1, which presents objective data, and Table 5.2, which presents subjective data, are the same.

Tables 5.1 and 5.2 were adapted from *Journal of Food Science*, Vol. 66, No. 6, p. 787, 2001 and *Journal of Food Science*, Vol. 67, No. 5, p. 1967, 2002. Copyright by Institute of Food Technologists. With permission.

Note:

- No lines are within the table itself.

- Title is not indented.

- There are no units of measure in a subjective table.

- Footnote the number index that is used to denote the subjective characteristic (for instance, if a scale of 1–5 is used, indicate whether 5 is the highest or lowest value on the scale).

- Subjective evaluations should be put in a separate table from objective evaluations.

For recording data in laboratory books, each trial, as well as the average, needs to be recorded. An example of a style that would be suitable is:

	Variable
	0.89
Treatment	0.86
	0.83
Average	0.86

TABLE 5.2

Effects of Edible Coatings on Sensory Attributes of Baby Carrots During Storage*

Time	Treat[+]	White Discoloration	Orange Intensity	Crispness	Sweetness	Bitterness	Fresh Aroma	Fresh Flavor	Slipperiness
1st week	1	3.83^a	4.54^c	5.88^a	5.13^a	1.92^a	5.76^a	5.7^a	1.91^c
	2	1.68^{bc}	6.45^{ab}	5.92^a	4.52^{ab}	1.99^a	4.83^{ab}	5.25^a	5.14^b
	3	2.48^b	6.03^b	6.06^a	4.48^{ab}	2.23^a	4.43^b	4.89^{ab}	6.58^a
	4	0.92^c	7.27^a	6.13^a	4.03^b	2.98^a	3.83^b	3.89^b	7.57^a
2nd week	1	6.11^d	3.82^d	5.67^a	6.16^c	1.71^b	4.78^{ab}	4.82^{ab}	1.12^d
	2	2.36^b	6.49^{ab}	5.42^a	5.31^a	1.93^{ab}	5.68^a	5.71^a	3.10^c
	3	2.29^b	6.19^b	5.61^a	4.48^{ab}	1.95^{ab}	5.41^{ab}	4.66^{ab}	5.27^b
	4	1.40^c	7.44^a	6.01^a	4.41^{ab}	2.83^a	4.39^b	3.91^b	6.76^a
3rd week	1	6.19^d	3.52^c	5.69^a	5.40^{ac}	2.20^a	4.72^{ab}	4.33^{ab}	1.22^d
	2	2.41^b	6.44^{ab}	6.06^a	5.53^{ac}	2.17^a	4.51^b	6.39^a	3.23^c
	3	2.39^b	6.40^{ab}	6.02^a	5.24^a	2.07^a	5.72^a	5.08^{bc}	5.10^b
	4	0.90^c	7.39^a	5.59^a	4.90^{ab}	2.93^a	5.52^{ab}	5.54^a	6.62^a

[a-d] Mean values with different superscript letters in the same column differ ($p<0.05$).

* Sensory scale: 0 = none, 10 = intense.

[+] Treatment: 1 = control, 2 = xanthan gum alone, 3 = xanthan gum/vitamin E, 4 = xanthan gum/calcium.

For the individual projects, replications will be performed over three weeks. It is inefficient to make a new table for each week, so tables should be prepared in advance in the laboratory notebook. An example of a table that might be appropriate is illustrated by Table 5.3.

In the final report, report data as means with standard deviations.

TABLE 5.3

Example of a Table for Collecting Data over Multiple Weeks

	Replication			
Treatment	I	II	III	Average

■■ GRAPHS FROM THE *JOURNAL OF FOOD SCIENCE**

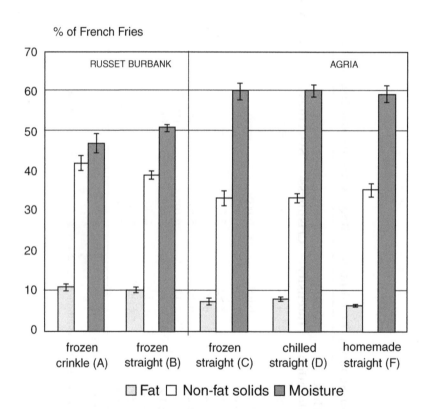

Figure 5.1 Composition of the five different types of French fries after finish-frying (180°C, 3.5 min). The lines on the bars are standard deviations.

* From Vol. 66, No. 6, p. 905, 2001; Vol. 66, No. 7, p. 930, 2001; and Vol. 66, No. 5, p. 656, 2001. Copyright by Institute of Food Technologists.

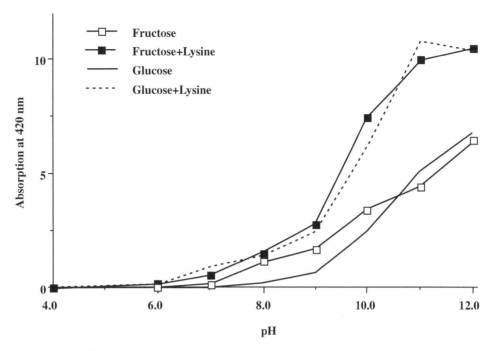

Figure 5.2 Effect of pH on browning development in aqueous fructose, glucose, fructose–lysine and glucose–lysine model systems heated to 100°C for 60 minutes.

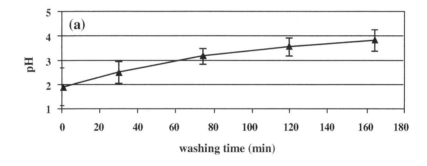

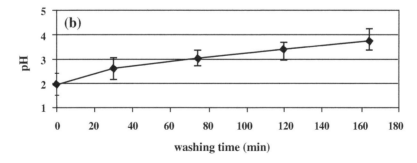

Figure 5.3 pH changes of pectin gel during washing with aqueous alcoholic solution: (a) aqueous–ethanol solution; (b) aqueous–2-propanol solution.

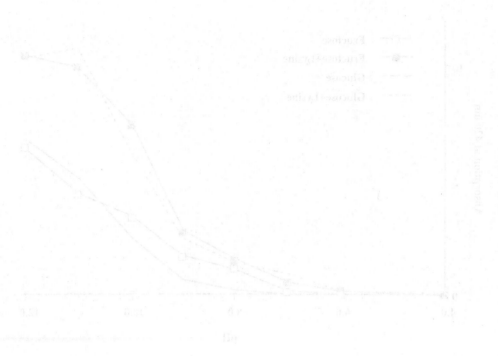

6 STYLE GUIDE FOR RESEARCH PAPERS

This chapter presents the information that is provided by the Institute of Food Technologists (IFT) for authors who wish to publish papers in the *Journal of Food Science* (JFS) or IFT's other scientific journals.

■ MISSION OF IFT SCIENTIFIC JOURNALS

The Institute of Food Technologists (IFT) publishes scientific journals to provide its members with scientific information that is important and of current interest. This is done in accord with the highest standards of professional ethics. Research articles serve to convey the results of original work that has a clear relationship to human foods. Review articles serve to convey in-depth, interpretive coverage of topics of current importance. Acceptability of articles for publication is carefully considered, with quality of the science, appropriateness, and importance weighing heavily in the final decision.

■ GENERAL EDITORIAL POLICIES

1. Authorship Criteria and Author Responsibilities

IFT is proud of the high quality of research reported in its journals and is dedicated to maintaining a high level of professionalism. However, because unprofessional behavior, either intentional or unintentional, has been known to occur, we remind authors of their obligations when submitting manuscripts for publication.

2. Authorship Criteria

Authorship is restricted to those who:

> Have contributed substantially to one or more of the following aspects of the work: conception, planning, execution, writing, interpretation, and statistical analysis
> Are willing to assume public responsibility for the validity of the work

> Membership in the Institute of Food Technologists is not a prerequisite for consideration of manuscripts for publication.

3. Exclusivity of Work

The corresponding author must verify, on behalf of all authors (if more than one), that neither this manuscript nor one with substantially similar content has been published, accepted for publication, or is being considered for publication elsewhere, except as described in an attachment.

4. Disclosure Requirements

With manuscript submission, authors must disclose affiliation or involvement, either direct or indirect, with any organization or entity with a direct financial interest in the subject matter or materials discussed in the manuscript (e.g., employment, consultancies, stock ownership, grants, patents received or pending, royalties, honoraria, expert testimony). Specifics of the disclosure will remain confidential. If deemed appropriate by the Editor, a general statement regarding disclosure will be included in the Acknowledgment section. Authors must disclose, in the Acknowledgment section of the manuscript, all sources of support for the work, both financial and material.

5. Copyright

Copyright to published manuscripts becomes the sole property of the Institute of Food Technologists. The corresponding author will be asked to sign a Copyright Transfer Agreement on behalf of all authors. In instances where the work cannot be copyrighted (works authored solely by government employees as part of their employment duties), this requirement is waived.

Reproduction of all or a portion of a JFS article by anyone, including authors, is prohibited, unless permission is received from IFT. Authors have the right to reproduce extracts from the paper with proper acknowledgment and retain the right to any patentable subject material that might be contained in the article. Requests for permission to reproduce material should be made in writing to: Director of Publications, Institute of Food Technologists, 221 N. LaSalle Street, Suite 300, Chicago, Illinois 60601–1291, U.S.A.

6. Disclaimer

Opinions expressed in articles published in this journal are those of the author(s) and do not necessarily represent opinions of the IFT. IFT does not guarantee the appropriateness for any purpose, of any method, product, process, or device described or identified in an article. Trade names, when used, are only for identification and do not constitute endorsement by IFT.

7. Criteria for Manuscript Acceptance

Manuscript acceptability is based primarily on quality of the work (clarity of objectives; originality; appropriate experimental design and methods; appropriate statistical analysis; depth of the investigation; substance of the results; thoroughness with which the results are discussed; appropriate conclusions), and appropriateness and importance of the topic.

8. Page Charges

A charge of $70 (subject to change) per printed page is currently assessed for all manuscripts on original research that are submitted electronically through Manuscript Central (see below). Submission of manuscripts in paper form is discouraged and may be assessed an additional charge. When payment is possible only from personal funds, and this means of payment would impose undue financial hardship, a request for waiver of this charge can be made, provided this is done prior to publication. In this instance, a statement certifying that the author's employer(s) is unable to pay because of financial distress, and that the author cannot personally pay because this would impose an undue financial burden, signed by both the author and the employer, should be sent to the Senior Editor at the address listed below.

Concise Reviews and Hypothesis Papers are exempt from page charges, provided the Scientific Editor (Daryl B. Lund dlund@cals.wisc.edu) is consulted and issues an invitation in advance of submission.

9. Reprints

Following acceptance of a paper and prior to publication, the author will be given the opportunity to purchase reprints. An order form and rate schedule will be included with the manuscript's page proof.

10. Permission to Publish

If the paper has been presented at a meeting of an organization other than IFT, the author must certify that he/she has freedom to offer it to IFT for publication.

■ JOURNAL SECTIONS

Authors are asked to indicate the desired section for their manuscript when submitting the paper. Choose among:

1. JFS — Concise Reviews and Hypotheses in Food Science. SCIENTIFIC EDITOR: DARYL B. LUND. Coverage of all aspects of food science, including safety and nutrition. Reviews should be 15 to 50 typewritten pages (including tables, figures, and references), should provide in-depth coverage of a narrowly defined topic, and should embody careful evaluation (weaknesses, strengths, explanation of discrepancies in results among similar studies) of all pertinent studies, so that insightful interpretations and conclusions can be presented. Hypothesis papers are welcome. These are especially appropriate in pioneering areas of research or important areas that are afflicted by scientific controversy.

2. JFS — Food Chemistry and Toxicology. SCIENTIFIC EDITOR: STEPHEN L. TAYLOR. Coverage of original research on degradative and preservative reactions, toxicology, functional properties, post-harvest physiology of plants, muscle biology, analytical procedures, and composition.

3. JFS — Food Engineering and Physical Properties. SCIENTIFIC EDITOR: ROMEO T. TOLEDO. Coverage of original research on quantitative aspects of unit operations associated with food preservation/processing and food waste recovery, with emphasis on systems design and analysis, modeling, simulation, optimization, physical properties measurement and instrumentation, thermodynamic relationships, sensors and automation, and materials science, including surface properties and interactions, rheology, mass transport properties, water activity, and glass transitions.

4. JFS — Food Microbiology and Safety. SCIENTIFIC EDITOR: ELLIOT T. RYSER. Coverage of original research on foodborne pathogens, pathogenesis, risk assessment, spoilage, fermentation, preservation, microbial growth/inactivation, biotechnology, and methods.

5. JFS — Sensory and Nutritive Qualities of Food. SCIENTIFIC EDITOR: TUNG-CHING LEE. Coverage of original research on flavor, color, and texture assessment, both quantitative and subjective; nutritional properties; nutraceuticals; and quality attributes as influenced by processing/storage/packaging.

6. Web site — *Comprehensive Reviews in Food Science and Safety*. SCIENTIFIC EDITOR: DAVID L. LINEBACK. May deal with any aspect of food science, including safety and nutrition, that is of widespread current interest. Should provide in-depth coverage of a narrowly defined topic, and should embody careful evaluation (weaknesses, strengths, explanation of discrepancies in results among similar studies) of all pertinent studies, so that insightful, integrative interpretations, summaries, and conclusions can be presented. Before preparing a manuscript, the author should submit: a proposed title; a short statement describing the importance of the topic and how the presentation will advance the field of food science (for unsolicited papers only), and a one-page outline. Following agreement between the author and Scientific Editor with respect to title, author statement, and outline, the author will receive an invitation to prepare a manuscript.

7. Web site — *Journal of Food Science Education* (JFSE). SCIENTIFIC EDITOR: WAYNE IWAOKA; INTERNATIONAL SCIENTIFIC EDITOR: ALBERT J. McGILL. Publishes information about the teaching of food science and technology and serves as an information vehicle to instructors in food science at various educational levels. Appropriate information includes: results of original

research studies dealing with instructional methods, news about educational systems for teaching food science, information about the interaction among teaching methods, and content of food science courses. When submitting a manuscript through Manuscript Central, please indicate your section choice in the "Comments to Editor-in-Chief" window. Specify one of the following: original research, review, innovative laboratory exercise or demonstration, classroom technique, or Letter to Editor.

■ MANUSCRIPT REQUIREMENTS

Unless otherwise stipulated, the style and format of manuscripts submitted to *JFS* and the two Web site-based journals should follow the *Scientific Style and Format: The CBE Manual for Authors, Editors and Publishers*, 1994, 6th ed. (Council of Biology Editors, Cambridge University Press, New York). For convenience, refer to articles in the latest issue of the journal for details or contact the JFS Editorial Office with your questions. CLICK HERE for Supplementary Instructions for preparing manuscripts on special topics (flavor, fruits and vegetables, nutrition, engineering, etc.).

Use the English language (American spelling and usage) and the SI system (Système International Unités, often referred to as "International Units") for measurements and units.

All manuscripts should be submitted electronically through Manuscript Central. Details provided at the end of this document.

Working Template for Research Papers

Use this working template as a visual guideline. Simply remove the guides and fill in the appropriate information. (Word 6.0 format.)

Manuscripts on Original Research
Manuscripts on original research should include the following elements.

Title Page, as p. 1.
Include:

Full title (be concise)

Name(s) of author(s) and author affiliation(s) with complete address(es)

Contact information for the corresponding author, including full name, complete mailing address, telephone, fax, and e-mail address

Short version of title (less than 40 letters and spaces)

Choice of the journal section in which you would like your article to appear, choosing from those listed above

Previous address(es) of author(s) if research was conducted at a place different from current affiliation

Manuscript Central will indicate where this information should be entered.

Abstract Page, as p. 2.
Include:

An abstract not exceeding 110 words; all acronyms and abbreviations defined; no references cited. State what was done, how it was done, major results, and conclusions. Five key words for indexing purposes.

Manuscript Central will indicate where this information should be entered.

Introduction

In two pages or less, review pertinent work, cite key references, explain importance of the research, and state objectives of your work.

Materials and Methods

Provide sufficient detail so work can be repeated. Describe new methods in detail; accepted methods briefly with references. Use subheadings as needed for clarity.

USE OF TRADE NAMES

Trade names are to be avoided in defining products whenever possible. If naming a product trade name cannot be avoided, the trade names of other like products also should be mentioned, and first use should be accompanied by the superscript symbol ™ or ®, followed in parentheses by the owner's name. If a product trade name is used, it is imperative that the product be described in sufficient detail so the nature of the product will be understood by professionally trained readers. Do not use trade names in titles.

USE OF ABBREVIATIONS AND ACRONYMS

At first use in the text use abbreviated term, followed by abbreviation or acronym in parentheses. Do not use abbreviations and acronyms in titles.

STATISTICAL ANALYSIS

If variation within a treatment (coefficient of variation, the standard deviation divided by the mean) is small (less than 10%) and difference among treatment means is large (greater than 3 standard deviations), it is not necessary to conduct a statistical analysis. If the data do not meet these criteria, appropriate statistical analysis must be conducted and reported.

Results and Discussion

Present and discuss results concisely using figures and tables as needed. Do not present the same information in figures and tables. Compare results to those previously reported, and clearly indicate what new information is contributed by the present study.

Conclusion

State conclusions (not a summary) briefly.

References

List only those references cited in the text. Required format of references is described below.

Acknowledgments

List sources of financial or material support and the names of individuals whose contributions were significant but not deserving of authorship. Acknowledgment of an employer's permission to publish will not be printed.

Appendix

This section is rarely needed in a research paper but can be added if deemed necessary (e.g., complicated calculations, detailed nomenclature).

Tables

Number each table with Arabic numerals. Place a descriptive caption at the top of each table. Print one table per page. Columns and their headings are usually (but not always) used to display the dependent variable(s) being presented in the table. Footnotes should be identified by lower-case letters appearing as superscripts in the body of the table and preceding the footnote below the table. The same data should not appear in both tables and figures. Tables (and images) must be submitted electronically in Manuscript Central.

Figures (Graphs, Charts, Line Drawings, Photographs)

Figures (images) must be submitted electronically in Manuscript Central. Type in the legend, with Arabic numbering, immediately below your image file reference in Manuscript Central.

Authors are responsible for obtaining permission to reproduce previously copyrighted illustrations. Proof or certification of permission to reproduce is required. Lettering, data lines, and symbols must be sufficiently large so as to be clearly visible when the figure is reduced to a size commonly used in the journal. When a color presentation is deemed necessary, please note this in the cover letter of the submission.

Review Manuscripts

Essential elements (described elsewhere except for "text") are title page, abstract, introduction, text, conclusion, references. Summary tables and figures dealing with key points should be used liberally. The review should begin with a statement describing the importance of the topic and the objectives of the review.

A standard format for headings in the text is not required, but headings and subheadings should be used whenever needed to improve the clarity and readability of the presentation. Authors are encouraged to consult with the editor-in-chief before preparing a review for consideration.

Hypothesis Papers

Essential elements are title page, abstract, text, conclusion, references. The paper should begin with a statement describing the objectives of the paper, be followed by a logical progression of ideas or concepts that provide a rationale for the hypothesis, and end with conclusions. Headings and subheadings in the text should be used at the author's discretion to improve clarity and readability of the presentation. Authors are encouraged to consult with the editor-in-chief before preparing a hypothesis paper for consideration.

Other Types of Papers for the Journal of Food Science Education

A standard format is not required. Choose sections and headings that are most appropriate for the type of data being presented.

■ REFERENCE FORMAT

Manuscripts intended for all sections of the journal and the Web site journals must follow the name-year reference format of the Council of Science Editors (formerly Council of Biology Editors). Cite only necessary publications. Primary rather than secondary references should be cited, when possible. It is acceptable to cite work that is "in press" (i.e., accepted but not yet published) with the pertinent year and volume number of the reference. Work that is "submitted" but not yet accepted should not be cited.

In Text

Cite publications in text with author name and year. Use "and others" rather than "et al." In parenthetical citations, do not separate author and year with a comma. Use commas to separate publications in different years by the same author. Semicolons separate citations of different authors. Cite two or more publications of different authors in chronological sequence, from earliest to latest. For example:

> The starch granules are normally elongated in the milk stage (Brown 1956).
> Smith and others (1994) reported growth......
> ... and work (Dawson and Briggs 1984, 1987) has shown that...
> ... and work (Dawson 1984; Briggs 1999) has shown that...

In References Section

List only those references cited in the text. References should be listed alphabetically by the first author's last name. Single author precedes same author with co-authors. Type references flush left as separate paragraphs (do not indent manually, let the text wrap). Use the following format (note use of periods):

Journal Articles
Author(s). Year. Article title. Journal title volume number (issue number): inclusive pages.

Example: Citation in text: (Smith and others 1999): Smith JB, Jones LB, Rackly KR. 1999. Maillard browning in apples. J Food Sci 64(4):512–518

Books
Author(s) or [editor(s)]. Year. Title. Place of publication: publisher name. Number of pages.

Example: Citation in text: (Spally and Morgan 1989): Spally MR, Morgan SS.1989. Methods of food analysis. 2nd ed. New York: Elsevier. 682 p.

Chapters
Author(s) of the chapter. Year. Title of the chapter. In: author(s) or editor(s). Title of the book. Edition or volume if relevant. Place of publication: publisher. Pages of the chapter.

Example: Citation in text: (Rich and Ellis 1998): Rich RQ, Ellis MT. 1998. Lipid oxidation in fish muscle. In: Moody JJ, Lasky, UV, editors. Lipid oxidation in food. 6th ed. New York: Pergamon. p 832–855.

Patents
Name of the inventor of the patented device or process; Company name, assignee. Date issued [year month day]. Title. Patent descriptor [including name of country issuing the patent and the patent number].

Example: Harred JF, Knight AR, McIntyre JS, inventors; Dow Chemical Co., assignee. 1972 Apr. 4. Epoxidation process. U.S. patent 3,654,317.

In Press Items
Identify as "Forthcoming," not "In press."

For journal abbreviations and other examples of reference formats please refer to articles in a 2000 or 2001 issue of this journal or contact the JFS Editorial Office at IFT.

■ EDITORIAL REVIEW AND PROCESSING

1. Peer Review

All submitted manuscripts are screened by the section's Scientific Editor for importance, substance, appropriateness for the journal, general scientific quality, and amount of new information provided. Those failing to meet current standards are rejected without further review. Those meeting these initial standards are sent to expert referees for peer review (except for Letters to the Editor). Referees' identities are not disclosed to the author. Author identities are disclosed to the referees. Reviewer comments are reviewed by an Associate Editor and he/she, often after allowing the author to make changes in response to the referee's comments, advises the Scientific Editor to either accept or reject the manuscript. The Scientific Editor informs the author of the final decision.

2. Accepted Manuscripts

The author(s) will be asked to review a fully laid out and copy-edited page proof. The author(s) is responsible for all statements appearing in the galley proofs. The author will be informed of the estimated date of publication.

3. Inquiries Regarding Status of the Manuscript

Direct inquiries to: Betsy Baird, Senior Editor; Institute of Food Technologists, 221 N. LaSalle St., Chicago, IL 60601; Telephone: (312) 782 8424; Fax: (312) 782 8348; E-mail: babaird@ift.org.

■ INSTRUCTIONS FOR SUBMITTING A MANUSCRIPT

1.   Electronic Submission

JFS now requires that all manuscripts (including the two new Web site–based journals) be submitted electronically via an Internet service called "Manuscript Central" before they can be considered for publication. (Authors who do not have access to the Internet may submit their manuscripts in paper form to the JFS Editorial Office.) Electronic submission, however, will speed the handling of your manuscript and allow you to monitor its status in the handling process at any time.

2.   Procedure

Go to your Internet browser and type in http://ift.manuscriptcentral.com to bring up the Log-in screen. Click on "Submission Instructions" to familiarize yourself with the procedure. Return to "Log-in" and click on either "Create an Account" (if you are a first-time user) or "Check for Existing Account" (if you have previously created an account).

3.   Create an Account

Click on this button, which will bring up the "Create Your User Account" screen; fill in the information requested (name, address, phone, fax, e-mail address; also a chosen User ID — this is necessary to access your account in the future). Once you have filled in this identifying information, return to the "Log-in" screen and log in. Be sure to make a note of your User ID since you will use it any time you access Manuscript Central. Do not create duplicate accounts.

4.   Manuscript Submission

Log in, click on "Author Center," then on "Submit a Draft Manuscript," and follow the prompts and instructions given on the screen until your manuscript (text, figures, tables) has been entered. Once you have finished, have double-checked for accuracy, and have clicked on the "Submit" button, you will receive a JFS manuscript number and an e-mail message verifying that your manuscript has been received and entered. You may then keep track of the status of your manuscript by logging on to Manuscript Central (http://ift.manuscriptcentral.com), where the status will be displayed in the Author Center, along with the name of the editor in charge of your review.

5.   If You Have a Technical Problem

Assistance with technical difficulties in submission is available from ScholarOne, Inc., the parent company of Manuscript Central. Click the "Get Help Now" button on your screen and consult "FAQs" (Frequently Asked Questions), or contact ScholarOne at telephone 804–817–2040, ext.167; fax 804–817–2020; or e-mail at Support@ScholarOne.com.

6.   Print Submission

If electronic submission is not available, print submission is acceptable. However, this method of submission will result in slower handling of your manuscript, your inability to monitor its status in the handling process, and, possibly, a higher page charge.
      Submit the following items:

   Cover letter. Identify the corresponding author and provide his/her full name and communication data (telephone number; postal, fax, and e-mail addresses). Indicate the journal and the section in which you desire to have your manuscript appear. If you believe some of your figures require color presentation, please indicate.

   Authorship statement. Include signed form for authorship criteria and responsibility, financial disclosure, and copyright.

Manuscript. Double-space all components of the manuscript except tables. Type on one side of $8\frac{1}{2}$-in. × 11-in. paper. Use 1-in. margins. Number all pages and lines.

Send an original manuscript (with original figures marked "ORIGINAL") and three photocopies. If photocopies of figures are not clear (for example, photomicrographs or gel electrophoresis photographs), please do not use photocopies. Staple each manuscript copy, including the original manuscript, in the upper left corner. Do not use paper clips or binders.

Disk. Include an IBM-formatted, $3\frac{1}{2}$-in. disk, containing the title page and abstract in WordPerfect® (version 8.0 or earlier) or in Microsoft® Word (version 97 or earlier). Please be sure the manuscript conforms to the JFS style as outlined above. A failure to use this style may result in delayed publication.

■ PRESUBMISSION CHECKLIST

(See "Instructions for Authors" for additional information.)

Cover letter and form
   Full contact information for the corresponding author (full name, address, phone, fax, e-mail)
   Choice for JFS journal section identified:
      Concise Reviews and Hypotheses in Food Science
      Food Chemistry and Toxicology
      Food Engineering and Physical Properties
      Food Microbiology and Safety
      Sensory and Nutritive Qualities of Food
   Statement on authorship criteria, financial disclosure, and potential conflict of interest, and copyright (completed form)
Manuscript
   Original manuscript (with original figures, marked "ORIGINAL")
   Three photocopies
   Line numbering
   Page numbering
   Each manuscript stapled in upper left corner (no paper clips or binders)
   Title page: Corresponding author's contact information on title page
   Abstract of less than 110 words
   Five key words
   Figure captions listed consecutively on a page separate from figures
Disk
   $3\frac{1}{2}$ in. IBM-formatted disk with title page and abstract in a Word Perfect (version 8.0 preferred) or Microsoft Word (Windows 98 preferred) file. (Submission of the title page and abstract by electronic means is an acceptable alternative.)

7 INDIVIDUAL PROJECT

Select a problem for your individual project as soon as possible, and have your topic and methods approved by the instructor. Develop a testable hypothesis with clearly identified independent and dependent variables. Apply the principles of the scientific method in approaching your problem.

Once the topic and methods are verbally approved, the plans should be formalized in writing for a grade. The following should be included in your proposal.

RESEARCH PROPOSAL

1. Title

2. Hypothesis and objectives

3. Background: Review the literature and establish what is known and what gaps remain to answer your question. Include a justification for studying your problem. Justify your dependent variables. Is there a logical relationship between the dependent and independent variables? Justify the methods you selected. Are they standard procedures? Give the purpose of your project.

4. Approach: Give methods — what you plan to do and how. Be specific. Include procedures and recipes and their source. Quantities of ingredients must be in metric units (e.g., grams, milliliters). If you are doing sensory evaluation, include an example of your scorecard and describe your panel. How do you plan to control variables other than the one tested — for example, variation within a food sample, temperature, mixing procedure, size of product, sample preparation required for testing, etc.? Show that you have thought through the problem.

5. Work plan: Plan each step — what you are going to do each week and the preparations required prior to the laboratory period. Plan to replicate as time allows, preferably three times.

6. Supplies needed: Turn in supply sheets and market orders with your proposal. List item and amount and when needed along with any specifications (brand, etc.). For some materials, it is beneficial to have the same lot or variety for each replicate, so enough should be ordered at one time for the entire project. Perishable items must be ordered as needed. Prepare a separate supply order form for each day that you wish to receive materials and date the order form for the date you wish to receive the items.

ORAL PRESENTATION

Your presentation should be 8 minutes long. Emphasize your results and discussion.

1. Background: Provide sufficient background to acquaint the audience with the problem being studied. Where appropriate, show the chemical structure or reaction under investigation.

2. Methods

    - Describe the general design including the independent and dependent variables being studied. If the design is complicated, show a flow chart.

    - Briefly outline methods; describe assays and explain the basics of what they measure using reactions or flow diagrams where appropriate.

3. Results: Describe results, preferably using figures. Tables and figures must have complete titles and axes/column labels with units, and variables must be identified. Do not expect the audience to remember what a treatment code stands for through a series of overheads. Figures and tables should report means and standard deviations for each treatment, including for sensory data. Be sure to report the number of replications used and define the scale for sensory data. If you use a bar graph for sensory data, have the highest value indicate the most desirable property.

4. Discussion

    - Interpret your results based on results from the literature or based on reasoned scientific explanations. You must cite at least one literature reference from an original research article during your oral presentation.

    - Summarize the take-home message. You should be sure to address the original hypothesis or question. For example, if your objective is to reduce the calories in a product, you must calculate the calorie reduction as well as report objective and sensory findings.

■ WRITTEN PRESENTATION

Your written report must be typed, spell-checked, and neat. Use a technical writing style. Avoid the use of first person, contractions, and colloquial and literary styles. Use proper grammar.

The title should be descriptive but not excessively long.

Your written report should include an abstract. An abstract is a one-paragraph summary of problem, methods, main findings, and take-home message.

Your written report should include the following sections:

1. Introduction: This section should state the problem being studied with sufficient background so that readers can fully understand the project. This will likely require a discussion of a chemical process learned in class such as oxidative rancidity (with reactions), starch gelatinization, gluten development, etc. This section may also include a review of methods available to test your dependent variable and an explanation for your selected approach. This section should include a statement of the purpose of the project including specific independent and dependent variables. This statement could come at the end of this section or, if the section is several pages long, it could come at the end of the first paragraph.

2. Methods

    - Subheadings will likely help.

    - Give your overall design then specific procedures/assays/formulas.

    - Include your sensory scorecard where appropriate.

    - Give sufficient detail so that the project could be repeated by someone else (e.g., include settings/probe for texture analyzer, any important temperature or pH controls, equipment type, sample preparation, etc.).

    - Discuss replications, randomization, and sampling.

3. Discussion

   - If calculations are used in creating data, sample calculations should be provided here or in the Appendix. If standard curves are used, include figure or correlation coefficient and p value.

   - Give scientific and literature-based explanations and potential sources of error in interpreting results. Discussions without sufficient citations from the literature will result in substantial point deductions.

   - Give your take-home message. The reader should be able to determine whether your project was successful.

   - Give suggestions for further work.

4. Results: Summarize data in tables and figures using complete titles that can be understood without reference to the text (including type of product if relevant). The text must refer to each table and figure, and tables and figures must be numbered sequentially.

5. References

   - Use the style of the *Journal of Food Science* in your reference list and in citations in the text. Avoid direct quotations of references. Paraphrase sources — do not plagiarize!

   - Limited use of general textbooks is acceptable. Emphasize original journal articles. The literature available on the selected topic should be well represented.

6. Appendix: Optional; if included, this section follows references.

■ SCORECARD FOR GRADE

Photocopy this scorecard to be turned in with your written report.

Individual Work

    Problem well thought out   _____

    Used class time to best advantage   _____

    Effort outside of class time   _____

    Careful planning for each day   _____

    TOTAL POINTS (out of   )   _____

Oral Report

    Justification   _____

    Background information (relevant, complete, logical)   _____

    Experimental design and methods rationale, description _____

    Presentation and interpretation of results   _____

    Conclusion supported by literature   _____

    Suggestions for future work   _____

    Effective visuals   _____

    Presentation   _____

    Response to questions   _____

    TOTAL POINTS (out of   )   _____

Written Report

    Abstract   _____

    Introduction   _____

    Justification   _____

    Materials and methods   _____

    Results   _____

    Discussion   _____

    References   _____

    TOTAL POINTS (out of   )   _____

Common Mistakes

    Spelling, grammar   _____

    Neatness and form   _____

    Incomplete sentences   _____

    Poor use of graphs, tables   _____

    Verbose introduction

    Analysis of results   _____

    Conclusions not supported by literature   _____

    Incorrect style of referencing:

        In text   _____

        In bibliography   _____

    Insufficient search of literature   _____

Additional Comments:

8 LABORATORY: SENSORY EVALUATION OF FOODS

Consumer acceptability of a food is dependent on many psychological and physiological factors. The following experiments have been designed to illustrate some principles of flavor as well as to acquaint the student with several types of tests used in sensory evaluations.

■ EXPERIMENT 1: THRESHOLD CONCENTRATIONS OF THE PRIMARY TASTES

Introduction and Objective

The four basic tastes are sweet, sour, bitter, and salty. Humans are variably sensitive to compounds provoking these tastes. The lowest concentration that can be recognized as one of the basic tastes is known as the recognition threshold. The objective of this exercise is to determine the approximate recognition threshold concentrations of salty, sour, sweet, and bitter solutions.

Materials

- Sucrose (mol wt 342.3 g/mol) — 0.01 M, 0.02 M, 0.04 M — 1000 ml of each

- Sodium chloride (mol wt 58.4 g/mol) — 0.01 M, 0.02 M, 0.04 M — 1000 ml of each

- Tartaric acid (mol wt 150.1 g/mol) — 0.0002 M, 0.0005 M, 0.0010 M — 1000 ml of each

- Caffeine (mol wt 1942 g/mol) — 0.0005 M, 0.001 M, 0.002 M — 1000 ml of each

- Applicator sticks

Completion time: 30 minutes
Complications: none

Procedure

Rinse the tongue with water, then apply a small amount of the solution with the lowest concentration in each series. Rinse the tongue again with water and apply the next most concentrated solution. Repeat until the taste is distinguishable and record that concentration.

■ EXPERIMENT 2: EFFECT OF TEMPERATURE ON TASTE

Introduction and Objective

All basic tastes have an interaction with temperature, that is to say they have a maximum intensity at a certain temperature (for a given concentration). At higher or lower temperatures the sensory impact is reduced. The objective of this exercise is to determine the effect of temperature on the sweetness of a sucrose solution.

Material

- Sucrose solution (100 g sucrose per liter) — 1 l divided into thirds and stored at the appropriate temperatures.

Completion time: 10 minutes
Complications: none

Procedure

Divide the sucrose solution into three portions. Bring one portion to 4°C, one to 30°C, and one to 49°C. Rank the three for sweetness. Rinse your mouth between tastes.

Most sweet   _____

Moderately sweet   _____

Least sweet   _____

■ EXPERIMENT 3: PERCEPTION OF PHENYLTHIOCARBAMIDE (PTC)

Introduction and Objective

Not all people can taste all basic tastes equally, or in some cases, at all. The ability to taste PTC (normally detected as bitter) is genetically inherited as a dominant trait. Roughly two-thirds to three-fourths of the population can taste PTC as bitter ("tasters"), while the remainder cannot ("nontasters"). The objective of this exercise is to determine whether you are a "taster" or a "nontaster".

Material

- Taste papers (PTC, Carolina Genetics, 17–4010).

Completion time: 5 minutes
Complications: none

Procedure

Apply paper to tongue, wait 30–60 seconds, remove paper, swallow saliva, and record response.

■ EXPERIMENT 4: COMPARISON OF SWEETNESS OF SUGARS

Introduction and Objective

Sugars are generally regarded as sweet, but not all sugars have the same degree of sweetness. Structure-function relationships cause sugars of different molecular structures and shapes to have different levels of sweetness. The objective of this exercise is to compare the relative sweetness of several mono- and disaccharides.

Materials

- Fructose — 50 g

- Glucose — 50 g

- Sucrose — 50 g

- Lactose — 50 g

- Applicator sticks

Completion time: 5 minutes
Complications: none

Procedure

Use applicator stick to apply sample to tongue. Wait 10–20 seconds, then compare and record the relative sweetness of the various sugars.

■ EXPERIMENT 5: IDENTIFICATION OF SAMPLES

Introduction and Objective

We interact with food by way of all our senses. Among the long-distance food-interacting senses are sight and smell. These senses inform our judgments as to the quality and acceptability of foodstuffs. Without the benefit of sight and smell, it can be very difficult to assess the quality or even the identity of foods. The objective of this exercise is to show the importance of sight and odor for the identification of a product, and, as a corollary, the contribution of aroma to food flavor.

Materials

- Five different juices (fruit and/or vegetable, e.g., apple, grape, apricot, grapefruit, tomato) — 1 l of each

- Blindfolds

- Cotton

Completion time: 30 minutes
Complications: none

Procedure

Plug nose with cotton, wear blindfold, and try to identify each sample by taste only. Remove plug, taste the same five liquids again, and identify.

Question

Which samples were correctly identified by taste alone? By taste and odor?

■ EXPERIMENT 6A: DIFFERENCE TESTING

Introduction and Objective

Many methods can be used for testing foods. Some are fairly simple (for the panelist) "forced choice" methods such as paired comparison or the triangle test. Others are more complex, such as ranking or rating, category scaling, magnitude estimation, or hedonic rating. The objective of these exercises is to become acquainted with a variety of sensory difference testing methods that can be used with foods.

Material

- Apple juice solutions prepared according to the directions in the Appendix

Completion time: 30–45 minutes
Complications: none

Procedure

The test solution is apple juice to which has been added different amounts of 5% citric acid (see Appendix).

Concentrate on one aspect of flavor: tartness. *Taste*, do not drink, samples.

1. *Paired test*

    Simple difference: Are samples 545 and 390 of equal tartness or different?

2. *Directional difference*: Which sample (545 or 390) is more tart? _____

3. *Triangle or odd sample test*: Which sample (923, 517, or 886) differs from the other two in tartness? _____

4. *Ranking*: Rank the five samples (904, 792, 534, 459, 609) in descending order for tartness in the spaces provided.

    Most tart   _____
                     _____
                     _____
                     _____
    Least tart   _____

5. *Rating*: Rate samples (269, 109, 919) in descending order for tartness on the six-point scale below. The reference sample has a score of 4.

    (1) _____
    (2) _____
    (3) _____
    (4) _____   Reference sample (juice containing 2% of 5% citric acid)
    (5) _____
    (6) _____

■ EXPERIMENT 6B: DESCRIPTIVE TESTS

Category Scaling: Structured

Rate samples (512, 204, 843) for tartness against the descriptive terms below.

None   _____
Slight   _____
Moderate   _____
Strong   _____
Extreme   _____

Category Scaling: Unstructured

Rate samples (512, 204, 843) for tartness by drawing a vertical mark for each sample through the line and labeling it.

Not tart   Extremely tart

■ EXPERIMENT 6C: AFFECTIVE TESTS

Paired Preferences

Which sample (545 or 904) do you prefer? _____

Hedonic Rating Scale

Rate sample 843 by checking the appropriate box.

TABLE 8.1

Hedonic Rating Scale

☐	☐	☐	☐	☐	☐	☐	☐	☐
Dislike extremely	Dislike very much	Dislike moderately	Dislike slightly	Neither like nor dislike	Like slightly	Like moderately	Like very much	Like extremely

Questions

1. Did you correctly identify the more tart sample of apple juice in the paired test? What were your chances of guessing the right one?

2. Did you correctly identify the odd sample in the triangle test? If you were unable to distinguish among the three samples, how likely were you to select the odd sample by chance?

3. Which samples did you have out of order in the ranking test? How many paired comparisons had to be made to rank the five samples?

4. What was the concentration of acid in the reference sample in Experiment 6A, part 5? Did you identify correctly the test juice that had the same concentration of acid? Did you place the test juice with less acid below the reference sample? Did you place the test juice with more acid above the reference sample?

5. What is the purpose of the reference sample?

6. What factors might influence the position assigned to a particular sample in a descriptive test?

7. How could the unstructured test in Experiment 6B be quantified?

8. Which kind of sensory test can be used with consumers?

■ EXPERIMENT 7: ADAPTATION OF RECEPTORS

Introduction and Objective

The ability to taste or smell a given stimulus is mediated by taste or odor receptor cells and the biochemical reactions that go on inside them. The initial step in olfaction or gustation is binding of the molecule to be tasted or smelled to a receptor. The binding, action of the bound molecule, and release to get ready to repeat the cycle are a time-dependent series of events and thus can be saturated, resulting in so-called fatigue or adaptation. The objective of this exercise is to illustrate the adaptation of gustatory and olfactory receptors. Adaptation may be defined as the loss of or change in sensitivity to a given stimulus as a result of continuous exposure to that stimulus or a similar one.

Materials

- 3% sodium chloride solution — 500 ml
- Coffee concentrate — 500 ml
- Stopwatch

Completion time: 10 min
Complications: none

Procedure

1. Place a small amount of sodium chloride solution in the mouth and note the time at which the sensation of saltiness has subsided.

2. Place a beaker of warm coffee concentrate (70°C) about 3 inches from the nostrils, inhale strongly, and exhale. Record the intensity value of odor on a scale from 0 to 5, with 0 = no odor, 5 = strong odor. Repeat the inhalation and exhalation ten times. Each time record the odor intensity. Continue the inhalation and exhalation, and record the total time (min) required for complete adaptation.

LABORATORY:
9 OBJECTIVE EVALUATION OF FOODS

This laboratory period is designed to acquaint the student with some of the objective methods for evaluating the constituents of foods either through demonstration or through participation. This exposure should aid the student in subsequent laboratories and in planning individual research projects. Some methods that will be used in the laboratory regarding dispersion of matter will not be included in this laboratory. Directions for use of individual pieces of equipment are given in the Equipment Guide section of this manual.

■ TEXTURE

1. Tenderness

   - Textures analyzer — Test any kind of food.

   - Shear press — Test vegetables and meat.

   - Compressometer/penetrometer — Test baked products, cheese, gels.

   - Shortometer — Test crackers, pastry.

2. Cell structure

   - Photocopy — Test baked products.

3. Volume

   - Seed displacement — Test baked products.

   - Compensating polar planimeter — Test muffin.

   - pH meter — Test fruit juices.

4. Color

   - Hunter colorimeter or Photovolt reflectance meter — Test three varieties of milk such as whole, 2%, skim.

10 LABORATORY: PHYSICAL PROPERTIES OF FOODS

Some of the properties exhibited by foods and food components that will be studied in this laboratory include water activity, viscosity, specific gravity, and refractive index.

■ WATER ACTIVITY

The water activity (a_w) of a food or food ingredient is often more directly related to quality in terms of (1) shelf life stability as affected by chemical, enzymatic, and microbiological changes and (2) compatibility with other foods, formulation, and packaging requirements than is the moisture content of the food itself. The a_w directly or indirectly affects or is related to texture, appearance, aroma, taste, freeze-thaw stability, and the microbiological, chemical, and many other objective and subjective characteristics of food.

Water activity is defined as p/p_o, where p is the partial pressure of water over the food sample and p_o is the vapor pressure of pure water at the same temperature. Water activity may also be defined as the Equilibrium Relative Humidity (ERH) divided by 100.

$$a_w = p/p_o = ERH/100$$

The a_w of a food product may be measured absolutely by instruments such as the Water Activity System meter (see the Equipment Guide in the back of this book) or adjusted to a specific a_w by equilibrating food samples in chambers over saturated solutions (which produce a known ERH in the chamber) and testing them using objective texture measurements.

■ VISCOSITY

All fluids possess definite resistance to flow, and many solids show a gradual yielding to forces tending to change their form. This property is due to the internal friction of matter and is called viscosity. Viscosity measurements are useful in determining the degree of hydrolysis of starches, pectins, and proteins; the amount of an additive to be added to a food product (such as a gelling agent or emulsifier); the extent of a process affected by heat such as protein denaturation; and the moisture content of products such as honey. Viscometers are available that allow the determination of absolute viscosity, which is expressed in milliPascal seconds (mPs). An older unit for viscosity is centipoises (cP). One poise equals one dyne-second per cm^2. One mPs equals one cP. Often an experimenter is interested in the relative viscosity of a solution compared to a control such as water. This information can be provided easily by timing the flow of the solutions through a tube such as a jelmeter or a pipette.

Fluids may be classified as Newtonian or non-Newtonian. The flow of Newtonian fluids is characterized by a velocity of flow that is directly proportional to the force applied. Corn syrup is a true solution and should exhibit Newtonian viscosity with increased applied force. Water is also Newtonian in nature. Catsup is Bingham plastic in nature and exhibits non-Newtonian viscosity, which means it exhibits a changing apparent viscosity with increased applied force. The terms apparent viscosity and consistency can be used interchangeably for non-Newtonian fluids. Most food systems exhibit non-Newtonian viscosity. Both Newtonian and non-Newtonian fluids exhibit a decrease in viscosity with an increase in temperature (but certain bacterial gums such as xanthan and gellan show little change in viscosity over a wide temperature range).

▄ SPECIFIC GRAVITY

Density, the mass per unit volume at a specified temperature, is a physical property that can be used to identify foods. A related parameter, specific gravity, is often used instead of density. Specific gravity is the ratio of the density of a substance at a specified temperature to that of water at 4°C. The specific gravity of water at 4°C is 1.0. The density of water at 20°C is 0.99823 g/cm^3, which is close enough to unity that the temperature of the water will not need to be at 4°C for most uses. Specific gravity measurements are used to determine adulteration, quality, and composition of food such as the water or butterfat content of milk, the syrup concentration of canned fruits or juices, and the alcohol content of beverages.

Specific gravity of liquids can be measured with a hydrometer. The function of the hydrometer is based on Archimedes' principle that a solid suspended in a liquid will be buoyed up by a force equal to the weight of the liquid displaced. The lower the density of the sample, the lower the hydrometer will sink. Saccharometers are hydrometers graduated to indicate percentage of sucrose by weight (degrees Brix). The refractometer provides an alternative means of measuring the soluble solids of sugar syrups and fruit products and is based on the index of refraction of the solution.

▄ EXPERIMENT 1: WATER ACTIVITY

Introduction and Objective

Water activity is of critical importance to many qualities of foods such as texture, appearance, and shelf life. Foods can be adjusted to various water activities by equilibration of the food in a closed container (e.g., a desiccator) over concentrated solutions of various salts (see Table 10.1). The objective of this exercise is to demonstrate the effect of varying relative humidities (and thus, a_w) on the texture and visual sensory quality attributes of foods and to determine their approximate a_w.

TABLE 10.1

Saturated Solutions for Humidity Chambers

Relative Humidity (%)	a_w	Compound	Concentration to Make Saturated Solution at Ambient Temperature
7.0	0.070	Sodium hydroxide: NaOH	120 g/100 ml H_2O
22.5	0.225	Potassium acetate: $KC_2H_3O_2$	253 g/100 ml H_2O
42.8	0.428	Potassium carbonate: K_2CO_3	112 g/100 ml H_2O
56.0	0.560	Calcium nitrate $Ca(NO_3)_2 \cdot 4H_2O$	121 g/100 ml H_2O
81.8	0.818	Ammonium sulfate $(NH_4)_2SO_4$	80 g/100 ml H_2O

Note: Use distilled or deionized water to prepare solutions. Solutions may be prepared hot; however, cool them sufficiently prior to placing them in chambers. *Caution:* The NaOH solution is especially hazardous. Treat all solutions with extreme care.

Methods and Materials

Prepare in five humidity chambers (desiccators) the saturated solutions shown in Table 10.1.

Place representative samples (at least five of each) in 2-ounce plastic soufflé cups and store in evacuated humidity chambers for 2 weeks. Compare to fresh controls.

Evaluate the foods listed in Table 10.2 utilizing the techniques suggested there.

TABLE 10.2

Food Products and Methods for Their Evaluation After Storage at Different Humidities

Food Product	Evaluation Techniques
Soda crackers	Shortometer; texture analyzer — knife probe
Fig Newtons	Penetrometer — needle point; texture analyzer — puncture probe
Ripened cheddar cheese	Penetrometer — needle point + 50 g weight; texture analyzer — puncture probe
Cream cheese	Penetrometer — cone; texture analyzer — cone probe
Carrots or celery	Warner Bratzler shear press; texture analyzer — knife probe
Hard candies	Word description; texture analyzer — cylinder probe, tension mode

Determine the actual a_w of the stored crackers and cream cheese using the Water Activity System meter. *Note*: Do not eat the food products.

Completion time: 1 hour

Complications: Foods in desiccators should be prepared two weeks ahead of time and stored in a refrigerator to prevent mold growth.

Evaluation and Discussion Questions:

1. Graph the objective data vs. a_w. Determine the approximate a_w for optimal quality (fresh product) for each food product.

2. Was the a_w determined on the Water Activity System meter close to the value expected from the relative humidity of the saturated solutions? What conditions might account for discrepancies?

3. Are a_w and product moisture directly related? Discuss.

4. What are the ramifications of packaging on product shelf life in relationship to a_w?

5. When a cheese and cracker snack food is made for retail distribution, why do the a_w of the cheese and that of the cracker have to be the same? How does the cheese still maintain its "soft" texture? (*Note:* consider humectants.)

6. Suggest other tests that could objectively or subjectively evaluate the effect of a_w on foods.

For more on a_w, see:

1. Rockland, L.B., Saturated salt solutions for static control of relative humidity between 5° and 40°C., *Anal. Chem.*, 32, 1375, 1960.

2. Karrel, M., Water activity and food preservation, in *Principles of Food Science — Physical Principles of Food Preservation*, Marcel Dekker, New York, 1975, part II, p. 253.

■ EXPERIMENT 2: VISCOSITY

Introduction and Objective

Many foods can be described as viscous fluids. Viscosity (thickness) is an important property of foods and will determine the taster's sensory reaction to them. However, viscosity is not the only important factor; the type of flow, Newtonian or non-Newtonian, is also important in determining foods' ultimate sensory characteristics. The objectives of this exercise are to illustrate the operation of a Bostwick consistometer and a linespread apparatus for determining relative viscosity and the effect of temperature on viscosity and to illustrate the operation of a Brookfield viscometer for studying the influence of applied force on apparent viscosity.

Materials

- Catsup — 1 l

- Corn syrup — 1 l

- Consistometer

- Linespread apparatus

- Brookfield viscometer

- Two 600-ml beakers

Completion time: 30 minutes
Complications: none

Procedure

1. Fill two beakers with corn syrup. Bring one beaker to 10°C and the other to 20°C. Determine the relative viscosity of each using the linespread apparatus. Refer to Equipment Guide section for use. Time flow for 1 min.

2. Repeat Step 1 using catsup instead of corn syrup and determine consistency using the consistometer instead of the linespread apparatus. Time flow for 2 min.

3. Fill one 600-ml beaker with corn syrup and one 600-ml beaker with catsup. Determine the viscosity of each at 6, 12, 30, and 60 r/min with a Brookfield viscometer using spindle number 4. Refer to the Equipment Guide section for use of the Brookfield viscometer.

4. Graph the data.

Questions

1. What factors influence the apparent viscosity of a fluid?

2. How is the Brookfield viscometer used to determine whether a fluid is Newtonian or non-Newtonian?

■ EXPERIMENT 3: SPECIFIC GRAVITY AND REFRACTIVE INDEX

Introduction and Objective

Many Newtonian fluids are fairly dilute solutions of relatively low molecular weight solutes. Examples include sucrose dissolved in water to make a syrup or sugar in fruit juices. Alternatively, in emulsions containing various amounts of dispersed fat/oil, the specific gravity may differ. The objectives of this exercise are to illustrate the use of hydrometers and refractometers in determining the absolute or relative concentration of a substance in a liquid.

Materials

- Hydrometer
- Saccharometer
- Refractometer
- Whole milk — 250 ml
- Skim milk — 250 ml
- 17% (w/w) sucrose solution (Solution A) — 250 ml
- 27% (w/w) sucrose solution (Solution B) — 250 ml
- Hydrometer with scale in the range of 1 to 2 or a lactometer
- Saccharometer — Brix scale 10 to 50%

Completion time: 30 minutes
Complications: none

Procedure

1. Bring whole milk and skim milk to the temperature at which the hydrometer is calibrated. Transfer samples to appropriate cylinders. Gently lower the clean, dry hydrometer into each sample. Read scale when bubbles are gone and hydrometer is at rest. Further information on use of the various types of hydrometers may be found in the Equipment Guide section.

2. Use the saccharometer to determine the percentage sucrose in a 17% sugar solution and a 27% sugar solution.

3. Use the refractometer to determine the soluble solid content of the same sucrose solutions as in Step 2. Refer to the Equipment Guide section for use of the refractometer. Compare the results obtained with the saccharometer and the refractometer.

Questions

1. What is the effect of butterfat concentration on the specific gravity of milk?

2. What kind of dispersion is milk?

3. When would it be desirable to use a saccharometer vs. a refractometer?

11 LABORATORY: DISPERSION OF MATTER

A food dispersion is a system consisting of one or more dispersed or discontinuous phases in a continuous phase. In food systems, the continuous phase is usually either water or an edible oil. Dispersions can be classified on the basis of particle size. A true solution is a one-phase system with the molecules having dimensions below 1 nm. Colloidal dispersions consist of two or more phases, the dispersed phase consisting of particles ranging in size from 1 to 100 nm (10 to 1000 Å). A suspension has particles that have dimensions greater than this and that are subject to gravitational settling.

Particles in a true solution are sufficiently small that many particles can occupy a given volume. Solution properties that depend on the number of particles (but not on the identity of those particles) in a given volume are called colligative properties and include boiling point, freezing point, vapor pressure, and osmotic pressure. Two colligative properties, boiling point elevation and freezing point depression, will be studied in this laboratory. The mathematical expression is:

$$\Delta T = K_b \text{ or } K_f \times (\text{weight solute/weight solvent}) \times (1000/\text{mol wt of solute}) \times \text{\# of particles per molecule in solution}$$

where

K_b = the boiling point elevation constant, which is 0.51°C for each mole of solute in 1000 g of solvent when water is the solvent

K_f = the freezing point depression constant, which is −1.86°C for each mole of solute in 1000 g of solvent when water is the solvent

Dispersions can also be classified on the basis of the physical state of particles. The most common classifications on this basis in food systems are sols (solid dispersed in liquid), emulsions (liquid dispersed in liquid), and foams (gas dispersed in liquid). Many foods (fluid milk, salad dressings) consist of more than one dispersed phase in a continuous phase. Both emulsions and foams are stabilized by the presence of a surfactant, which lowers interfacial/surface tension, and by additives (emulsion stabilizers), which increase viscosity of the continuous phase.

■ EXPERIMENT 1: SOLUTIONS

Introduction and Objective

The solution colligative properties of boiling point elevation and freezing point depression are frequently seen in foods, for example in the endpoint cooking temperatures of candies and in the freezing points of frozen desserts, respectively. These colligative properties are a result of the interaction of compounds with water and depend on the number of particles present in solution but not on what those particles are. The objective of this exercise is to illustrate the effect of solutes on two colligative properties (freezing point and boiling point).

Materials

- Sucrose — 200 g

- NaCl (reagent grade)* — 40 g

Completion time: 1–1.5 hours

Complications: Be aware that these boiling solutions are *very* hot.

Procedure

1. Calibrate your thermometer by determining the boiling point of deionized (D.I.) water.

2. a. Prepare a salt solution with 30 g NaCl and 300 ml D.I. water in a 400 ml beaker.

   b. Place beaker on burner over an asbestos wire pad. Suspend thermometer in beaker using a thermometer stand. Begin heating to a rolling boil.

   c. Remove approximately 2 ml solution into premarked tubes at each of the following temperatures: 101, 105, 106°C. Do not undershoot temperatures. Calculate the expected salt concentrations (g solute/g solvent) at each boiling point using the formula on p. 49. Convert the concentration (wt. solute/wt. solvent) to percent by substituting wt. solute/wt. solvent in the formula with:

$$\frac{x}{x + 1} \times 100 = \%, \text{ where } x = \text{wt. solute/wt. solvent}$$

   d. Determine the percent soluble solids with a refractometer.

   e. Plot the boiling point vs. percent soluble solids measured on the refractometer and the calculated percent solids extrapolated from the figure on the same graph.

3. a. Prepare a sucrose solution with 30 g sucrose and 300 ml D.I. water in a 400-ml beaker.

   b. Place beaker on burner over an asbestos wire pad. Suspend thermometer in beaker using a thermometer stand. Begin heating to a rolling boil.

   c. Remove approximately 2 ml solution into premarked tubes at each of the following temperatures: 100.5, 101.5, 102°C. Do not undershoot temperatures.

   d. Determine the percent soluble solids with a refractometer.

4. a. Prepare a sucrose solution with 150 g sucrose and 300 ml water in a 400-ml beaker.

   b. Place beaker on burner over an asbestos wire pad. Suspend thermometer in beaker using a thermometer clamp and ring stand. Begin heating to a rolling boil.

   c. Remove approximately 2 ml solution into premarked tubes at each of the following temperatures: 103, 106.5, 112, 114, and 130°C. Do not undershoot temperatures; temperature will rise quickly between 112 and 114°C.

   d. Determine the percent soluble solids with a refractometer.

5. Calculate the expected sucrose concentration from Steps 3 and 4 and graph the boiling point vs. the expected and observed sucrose concentration on the same graph as described for salt in Step 2e.

---

* Table salt does not form a clear solution because it contains an anticaking agent.

6. a. Prepare 50 g of a 10% (w/w) solution of sucrose and 50 g of a 10% (w/w) solution of NaCl.

   b. Determine the freezing point of the 10% solutions by placing a thermometer in a test tube containing enough of the solution to cover the bulb of the thermometer. Place the test tube in a 4:1 ice to NaCl mixture and observe the temperature at which the solution solidifies.

Questions

1. Why do the calculated and observed concentrations differ for sucrose but not NaCl?

2. Why would you not be able to reach a boiling point higher than 107°C for NaCl?

3. What is the boiling point for a thread? Soft ball stage? Hard ball stage? Hard crack stage? How does the final boiling point relate to sucrose concentration and firmness of fudge or toffee?

4. Enumerate the characteristics of a true solution.

■ EXPERIMENT 2: EMULSIONS

Introduction and Objective

Emulsions, a mixture of one immiscible liquid in another, are common in food systems (salad dressings, mayonnaise, whole milk). Emulsions may be temporary (they separate in a few minutes) or permanent (they do not separate for months or longer). To achieve a permanent emulsion, an emulsifier is necessary. The function of an emulsifier is to reduce the interfacial tension between the water and oil phases and thus reduce the driving force for phase separation. The objective of this exercise is to demonstrate the effectiveness of various substances as emulsifying agents.

Materials

- Sudan red dye — 0.01 g
- Vegetable oil — 500 ml
- Electric blender or Sorvall Omni-Mixer
- Egg yolk — 2 ml
- Liquid detergent — 2 ml
- Microscope slides and cover glasses
- Lecithin — 2 g
- Polyoxyethylene sorbitan (Tween 40) — 5 ml
- Sucrose ester (Ryoto Sugar Ester S-170, HLB 1) — 2 g
- Sucrose ester (Ryoto Sugar Ester S-170, HLB 15) — 2 g
- Microscope
- Test tubes or small beakers

Completion time: 1–1.5 hours
Complications: none

Variations

See Table 11.1.

TABLE 11.1

Variations in Making Emulsions

Emulsifier	Oil (ml)	Water (ml)	Liquid for Dissolving Emulsifier
Oil/Water (O/W) Emulsions			
1. Control (no emulsifier)	10	40	—
2. Lecithin — 0.5 g	10	40	Oil
3. Egg yolk — 0.5 ml	10	40	Water
4. Detergent — 0.5 ml	10	40	Water
5. Bile — 0.5 g	10	40	Water
6. Polyoxyethylene sorbitan monopalmitate (Tween 40) — 0.5 g	10	40	Oil
7. Sucrose ester (Ryoto Sugar Ester S-170, HLB 1) — 0.5 g	10	40	Water
8. Sucrose ester (Ryoto Sugar Ester S-170, HLB 15) — 0.5 g	10	40	Oil
Water/Oil (W/O) Emulsions			
9. Control (no emulsifier)	40	10	—
10. Lecithin — 0.5 g	40	10	Oil
11. Egg yolk — 0.5 ml	40	10	Water
12. Detergent — 0.5 ml	40	10	Water
13. Bile — 0.5 g	40	10	Water
14. Polyoxyethylene sorbitan monopalmitate (Tween 40) — 0.5 g	40	10	Oil
15. Sucrose ester (Ryoto Sugar Ester S-170, HLB 1) — 0.5 g	40	10	Water
16. Sucrose ester (Ryoto Sugar Ester S-170, HLB 15) — 0.5 g	40	10	Oil

Procedure

1. Color oil with a small amount of the fat-soluble Sudan Red dye.

2. Mix emulsifier into liquid specified for your variation.

3. Put oil and water into sample cup of Sorvall Omni-Mixer or blender.

4. Mix for 30 seconds on speed 5 (Sorvall) or medium speed in the blender.

5. Pour into beakers or test tubes and observe.

6. Observe emulsions under a microscope to determine whether you made an oil/water (O/W) or a water/oil (W/O) emulsion.

Questions

1. Compare how the type of emulsifier affects emulsion formation and stability for W/O and O/W emulsions.

2. What chemical properties should a good emulsifier have?

3. How can emulsifiers be classified by their hydrophilic–lipophilic balance (HLB) number?

4. What kind of emulsion is cream? Butter? Margarine? Salad dressing?

5. Which type of emulsifier (high or low HLB value) would you select to manufacture margarine? Salad dressing?

6. Why are detergents good cleaning agents?

7. What is the role of bile in digestion and absorption of fats?

■ EXPERIMENT 3: FOAMING PROPERTIES OF PROTEINS

Introduction and Objective

Foams are a gas dissolved in a liquid and are common in food (whipped cream, foam on carbonated beverages). As dispersions of one phase in another, they are somewhat unstable in the same way that emulsions are unstable. However, some food ingredients, notably proteins, stabilize foams. The objectives of this exercise are to compare the foaming ability of various proteins, to investigate the mechanism of foam formation and stability, and to determine the effect of other chemical substances and temperature on protein foams.

Materials

- Egg albumin — 1 g

- Soy albumen (Mira-Foam 100, Gunther Products, Staley) — 1 g

- Whey protein concentrate — 1 g

- Oil — 1 ml

- Electric blender or Sorvall Omni-Mixer

- Sodium caseinate — 1 g

- Cornstarch — 1 g

- Sugar — 1 g

- 100-ml graduated cylinders

Completion time: 1–1.5 hours

Complications: Because this exercise requires that the samples be prepared and then observed over time, this experiment should be started first if the group doing it has more than one assigned experiment.

Procedure

1. Prepare 100 ml of the following dispersions in distilled water.

    a. 0.5% sodium caseinate

    b. 0.5% whey protein concentrate

    c. 0.5% egg albumin

    d. 0.5% soy albumen

    e. 0.5% soy albumen + 0.5% cornstarch

    f. 0.5% soy albumen + 0.5% sugar

    g. 0.5% soy albumen + 0.5% vegetable oil

    h. 0.5% soy albumen + 0.5% NaCl

2. Place 50 ml of solution a into sample cup of Sorvall Omni-Mixer, mix at speed 5 for 30 seconds, and place into 100-ml graduated cylinder. If using a blender, blend on medium speed for 30 seconds. Repeat with remainder of dispersion a. Continue for dispersions b–h. At the end of this step you should have two graduated cylinders containing each of the blended dispersions, a–h.

3. Place one set of the graduated cylinders containing dispersions a–h at room temperature and the other set in a 40°C water bath.

4. Measure foam volume at 0, 5, and 30 minutes after mixing.

Questions

1. Graph the loss of foam volume over time for each temperature for each dispersion.

2. Compare the volume and stability of the foams as a function of:

    a. Nature of protein present

    b. Type of ingredient in the presence of a protein

    c. Temperature

12 LABORATORY: LIPIDS

Lipids are widely distributed, and almost every natural food contains some quantity of them. Lipids are often added to foods during their preparation as a tenderizing agent, as one phase of an emulsion, as a method of transferring heat as in frying, or for flavor and richness in foods. Fats and oils (triglycerides; triacylglycerols) are characterized by their insolubility in water and solubility in organic solvents. One physical constant often used to identify fats and oils is specific gravity. The index of refraction of an oil is often used to indicate the degree of hydrogenation. Increases in the chain length or unsaturation of a fatty acid cause an increase in refractive index. Prevention of rancidity is a concern in the storage life of a fat or foods containing fats.

■ EXPERIMENT 1: ODORS AND PHYSICAL STATE OF LIPIDS AND FATTY ACIDS

Introduction and Objective

Compounds may or may not have odors, depending on their molecular weight. Molecular weight and other structural features (chain length, unsaturation) also affect fatty acid and lipid physical properties such as melting point. In the case of lipids, odors are principally produced by free fatty acids and, at least for low-molecular-weight fatty acids (C4–C10), the odors are unpleasant to say the least. The objective of this exercise is to demonstrate the odors and physical states of common food lipids and fatty acids.

Materials

- Lard — 10 g

- Hydrogenated vegetable oil — 10 g

- Refined corn or cottonseed oil — 10 ml

- Stearic or palmitic acid — 10 g

- Butyric or caproic acid (this need not be removed from the stock bottle; work in the fume hood with this fatty acid) — 1 ml

- Olive oil — 10 ml

- Butter — 10 g

- Margarine — 10 g

- Marine oils — 10 ml

- Sucrose esters — 10 g

- Sucrose polyesters — 10 g

Completion time: 15 minutes

Complications: It is not necessary to remove the top from the butyric acid container to smell it or, if you do, be sure to work in a fume hood.

Procedure

Place approximately 10 g of each lipid or fatty acid in 50-ml beakers, cover with aluminum foil, and bring to room temperature. Describe the odor of these fatty materials. Relate the observed odors to structure and degree of refinement or processing undergone.

▬ EXPERIMENT 2: SOLUBILITY, SPECIFIC GRAVITY, AND REFRACTIVE INDEX

Introduction and Objective

Lipids may be characterized in a number of ways. They are soluble in organic solvents but not in water. They have a specific gravity of less than 1.0. They may also be characterized by their refractive indexes, which may correlate to specific structural features such as unsaturation or chain length. The objective of this exercise is to illustrate some of these characteristics of lipids.

Materials

- Vegetable oils: corn, soybean, cottonseed, safflower, peanut, olive — 250 ml of each

- 20 ml chloroform

- 20 ml toluene

- 20 ml alcohol

- 250-ml graduated cylinders

- Hydrometer

- Refractometer

Completion time: 30–45 minutes

Complications: When working with organic solvents, be sure to work in a fume hood and use gloves.

Procedure

1. In a large test tube (25 × 200 mm), place 20 ml of water and 5 ml of vegetable oil. Shake vigorously and observe.

2. Place in each of three large test tubes (25 × 200 mm), respectively, 20 ml of toluene, 20 ml of alcohol, and 20 ml of chloroform. To each tube add 5 ml of vegetable oil. Shake vigorously and observe.

3. Nearly fill the graduated cylinder with vegetable oil at a temperature of 60°F. Introduce a hydrometer into the oil and determine the specific gravity of the oil.

4. Determine the refractive index of each of the oils. Refer to the Equipment Guide section for use of the refractometer.

Questions

1. In which solvents did the oil dissolve? Why?

2. What quality of oil is demonstrated by specific gravity?

3. Relate the structures of the various oils to their refractive indices.

■ EXPERIMENT 3: WATER-ABSORBING CAPACITY

Introduction and Objective

In certain food systems, fats must be mixed and remain mixed. The extent to which fats can absorb water is called the water-absorbing capacity and is important in food systems such as cakes. Fats differ in their ability to absorb water, largely based on differences in their composition. The objective of this exercise is to demonstrate the water-absorbing capacity of commercial fats.

Materials

-   Lard — 100 g

-   Hydrogenated shortening — 100 g

-   Margarine — 100 g

-   Burettes — 5

-   Soft margarine — 100 g

-   Butter — 100 g

Completion time: 1 hour
Complications: Add water slowly and gradually as you are beating the fats.

Procedure

Transfer 100 g of each lipid at room temperature to a small bowl of an electric mixer. While beating at slow speed, run water into the fat from a burette at a uniform rate (about 20 ml/min) until separation occurs. Record the volume of water taken up by 100 g of each fat. Graph the results.

Questions

1. What factors influence the emulsifying capacity of fats?

2. Why is the emulsifying capacity of a fat important?

■ EXPERIMENT 4: PLASTICITY OF FATS

Introduction and Objective

In addition to the incorporation of water into fats (Experiment 3), it is occasionally important to incorporate air into fats, as in the creaming of fat and sugar in cake mixes. This incorporation of air is largely responsible for the leavening that these products contain. The ability to incorporate air is related to fat crystal size (generally, the smaller the fat crystals, the more air that can be incorporated). The objective of this exercise is to illustrate the creaming ability of various fats.

Materials

- Lard — 100 g
- Hydrogenated shortening — 100 g
- Margarine — 100 g
- Soft margarine — 100 g
- Butter — 100 g
- Sugar — 750 g

Completion time: 1 hour
Complications: none

Procedure

Using the same fats as in Experiment 3 above, determine the relative amounts of air whipped in during the creaming operation as follows: Transfer 100 g of each fat to the small bowl of an electric beater. Add 150 g of fine sugar (granulated) gradually over a period of 2 min. Then continue to beat for another 3 min. Transfer the creamed mixture to a previously weighed measuring cup and determine the weight of 1 cup of creamed fat. Take the weight of a cup measure full of water to determine the specific gravities of the creamed fats. See Equipment Guide section for specific gravity. Graph your results.

Questions

1. What determines the creaming ability of a fat?
2. Why is the creaming ability of a fat important?

■ EXPERIMENT 5: FAT BLOOM IN CHOCOLATE

Introduction and Objective

Cocoa butter exists in a number of polymorphic forms (I–VI) of increasing melting points. Another system names these forms differently: $1 = \gamma$, II $= \alpha$, III and IV $= \beta'$, V $= \beta2$, and VI $= \beta1$. Fat bloom is a white or gray coating on the surface of chocolate that occurs when unstable forms of fat crystals melt, come to the surface, and recrystallize into the larger, more stable VI crystals. In order to prevent or delay bloom, it is necessary to set as much of the cocoa butter as possible in its stable V form initially and to avoid high storage temperatures. Small, unstable crystals are formed by rapid cooling and the presence of "seed" crystals. A predominance of the more stable crystals results if the partially crystallized chocolate is properly tempered (held at 32°C) to melt the unstable (I–IV) seed crystals that will begin to form, followed by moderate cooling. The objective of this exercise is to demonstrate how cooling rate and seeding can alter the crystal structure of chocolate in the tempering process.

Materials

Unsweetened chocolate — 2 oz
Aluminum pans

Completion time: 2 hours

Complications: If the group doing this experiment is also assigned another experiment, they should do this one first as it takes a considerable period of time with all the heating, cooling, reheating, and recooling.

Procedure

1. Melt two blocks (2 oz) of unsweetened chocolate in a 100-ml beaker. Use a water bath or oven to avoid overheating, but do not allow moisture to get into the chocolate. Chocolate is completely melted at 38°C.

2. Weigh 6 g of the melted chocolate into an aluminum dish and spread it evenly in a thin layer over the bottom. Place into a freezer for about 20 min. Remove and break into small pieces. Put it back into the freezer for 10 additional minutes to ensure adequate cooling.

3. After 30 min, adjust the remaining melted chocolate to 34°C. Place the frozen chocolate immediately into the warm chocolate. Stir the chocolate for 1 min or until pieces are melted and the mass is well mixed.

4. Pour half of the chocolate into an aluminum dish and refrigerate for 60 min.

5. Destroy any seed crystals in the remaining half by reheating the sample to about 40°C. Pour this into another aluminum dish and refrigerate for 60 min.

Questions

1. Compare the surface color and appearance of the samples and the unmelted square.

2. Describe why the specific treatments affected gloss as you observed it.

■ EXPERIMENT 6: OXIDATIVE RANCIDITY

Introduction and Objective

Because of their chemical structure, unsaturated fats and oils are subject to oxidative breakdown, so-called oxidative rancidity. This reaction is a free radical chain reaction that involves abstraction of a reactive allylic hydrogen from the fatty acid chain followed by a series of reactions with oxygen, rearrangements, and chain cleavage to produce odiferous compounds. In this experiment, carotene is used as a marker of fat rancidity reactions. The objective of this exercise is to study factors (light, temperature, antioxidants, prooxidants) that affect the rancidity of fats.

Materials

- Rendered pork fat— 50 g

- Carotene — 10 mg

- Chloroform — 1 ml

- Filter papers — 7-cm diameter discs

- Petri dishes with filter paper discs

- 0.01% $CuSO_4$ — 25 ml

- 0.001% BHA — 25 ml

- 0.5% hemoglobin — 25 ml

- Saturated salt solution — 25 ml

- Turnip greens extract — 80 ml

- Green onion tops extract — 80 ml

- White potatoes extract — 80 ml

Completion time: 2 hours

Complications: If the group or groups doing this experiment also have other experiments assigned, this one should be started first due to the length of time that must be allowed for the oxidation reaction to take place.

Procedure

To 50 g of rendered pork fat, add 10 mg carotene dissolved in a little chloroform. Using plastic forceps, bone forceps, or forceps coated in polyethylene, dip small filter papers (7-cm diameter is convenient) in the melted fat and allow to drain for 20 sec. Transfer to Petri dishes and treat as follows:

1. Effect of temperature and light on fat oxidation

   a. Cover the dish and store in the dark at room temperature.

   b. Cover the dish and store in the light (direct sunlight if possible).

   c. Cover the dish and store in the refrigerator.

   d. Cover the dish and store in an incubator at 60°C.

2. Antioxidants and prooxidants: For these experiments, saturate small filter discs with the test solutions (see below); place several discs of each test solution on a filter paper saturated with the carotene–lard mixture. Invert the bottom of the Petri dish containing the papers over the Petri top containing water, using a separate Petri dish for each variation. Store dishes in an incubator at 40°C. Solutions to be tested include:

   a. Water control

   b. Dilute copper solution (0.01% $CuSO_4$)

   c. Dilute hemoglobin solution (0.5%)

   d. Commercial antioxidant (0.001% BHA)

   e. Saturated salt solution

   f. Extracts prepared by heating 20 g chopped vegetable (turnip greens, green onion tops, white potato peel) with 80 ml H_2O to the boiling point. Decant and cool before using.

One cannot visualize fats going rancid. Carotene, which is a highly unsaturated hydrocarbon similar in structure to a fatty acid, turns from bright orange to colorless as it is oxidized. Therefore, the rate of bleaching observable with the carotene can be used as an index to the rate of oxidative rancidity of the fat. Compare the treatments for their influence on oxidative rancidity as determined by the rate of bleaching. Compare the odor of fat in bleached vs. nonbleached paper. Outline the events (with reactions) that occur as fats undergo oxidative rancidity. Discuss the results of each treatment using this outline.

13

LABORATORY: AMINO ACIDS, PROTEINS, AND MAILLARD BROWNING

Proteins are a major class of food constituents. These macromolecules have various properties that determine their behavior in foods. Some act as surfactants and have foam- and emulsion-stabilizing ability; some have great water binding properties that allow them to coagulate and form gels under certain conditions; and some are important for their enzymatic activity. The amino acid composition of proteins influences the functional qualities of individual proteins. Many foodstuffs possess distinctive color and odor characteristics as a result of reactions between amino groups and reducing compounds (Maillard reaction or Strecker degradation). Depending on the extent of formation, these pigments and odors may be desirable or undesirable. In addition, free amino acids influence taste sensations.

■ EXPERIMENT 1: MAILLARD REACTION

Introduction and Objective

Under certain conditions, reducing sugars may react with compounds bearing a free amino group and undergo a sequence of reactions known collectively as the Maillard reaction. As a part of this, alpha-dicarbonyl compounds produced in the Maillard reaction can react with amino acids and produce aromatic pyrazines. While a certain amount of browning and flavor generation is desirable in many foods, excess browning and aroma are undesirable. The objective of this exercise is to evaluate the aroma and color of heated amino acid–glucose solutions.

Materials

- D-Glucose — 50 mg
- L-Aspartic acid — 50 mg
- L-Lysine — 50 mg
- L-Phenylalanine — 50 mg
- L-Valine — 50 mg
- L-Methionine — 50 mg
- L-Leucine — 50 mg
- L-Proline — 50 mg
- L-Arginine — 50 mg

Completion time: 1.5 hours

Complications: This experiment takes a while so it should be started early in the lab period.

Procedure

1. To 50 mg of D-glucose in a test tube add 50 mg of an amino acid; add 0.5 ml of distilled water. Mix thoroughly.

2. Smell each mixture and record any sensations. Place a piece of heavy aluminum foil over each test tube top and heat the solutions in a water bath at 100°C for 45 minutes. Cool the contents to about 25°C in a water bath. Record the odor sensations for each solution (e.g., chocolate-like, potato-like, popcorn-like). Record the color as 0 = none, 1 = light yellow, 2 = deep yellow, 3 = brown. (*Note:* Color formation can be measured quantitatively if the solutions are diluted to 5 ml, except for arginine and lysine, which need to be diluted to 500 and 1000 ml, respectively.) Transfer the samples to colorimeter tubes and determine their absorbance at 400 nm. At 400 nm, the pigmentation or degree of browning is measured. What factors influence the degree of Maillard browning?

■ EXPERIMENT 2: QUALITATIVE TEST FOR PROTEIN

Introduction and Objective

Color reactions can be used for the qualitative, and in some cases the quantitative, analysis of proteins. These reactions are specific tests not for proteins but for certain structures commonly found in proteins. Therefore, ordinarily, one positive test is not conclusive proof of the presence of protein. To prove the presence of protein, several color tests should be run, and tests should be run that respond to a variety of characteristic groups of proteins. The objective of this exercise is to demonstrate color reactions of proteins.

Materials

- Variety of foods (e.g., egg, gelatin, milk, bread, flour, sugar, fat)

- Sodium hydroxide solution (1 N NaOH) — 10 ml

- Dilute $CuSO_4$ solution (0.01%) — 10 ml

- Concentrated nitric acid — 30 ml

- Potassium hydroxide solution (1 M) — 30 ml

- Lead acetate solution (1 M) — 30 ml

- Ninhydrin solution (0.35 g/100 ml ethanol) — 30 ml

- Test tubes

Completion time: 45 minutes

Complications: In the xanthoproteic test, be sure you always add the acid to the water dispersion of protein and not the other way around.

Procedure

Make suspensions of a small amount of each food in 5 ml water in each of five test tubes. Examine foods with each of the following tests:

1. Biuret test — Add strong alkali and then a few drops of dilute $CuSO_4$ solution to the protein solution. A violet color indicates a positive test. This test is specific for compounds with two or more peptide linkages; therefore, dipeptides do not give a positive test but all other polypeptides do. The structure of the Biuret complex is:

2. Xanthoproteic test — Carefully layer about 3 ml of concentrated nitric acid on the protein solution. Neutralize the acid with strong alkali; the yellow color changes to a burnt orange color if the test is positive. The test is positive for proteins containing amino acids with the benzene ring, such as tyrosine, phenylalanine, and tryptophan.

3. Reduced sulfur test — In a boiling water bath, boil the protein solution with a few milliliters KOH and then add a few milliliters lead acetate solution. A black precipitate (lead sulfide) is formed if sulfur-containing amino acids such as cystine and methionine are present.

$$\text{S-containing amino acid} + \text{KOH} \xrightarrow{\text{Pb(OAc)}_2} \text{PbS}$$

4. Ninhydrin test — Add a few milliliters ninhydrin (triketohydrindene hydrate) solution to a protein solution and heat the mixture to a boil in a boiling water bath. A lavender color appears on cooling if the compound contains at least one free amino group and one free carboxyl group, with the exception of proline and hydroxyproline, which give yellow products.

$$+ CO_2 \quad + \quad 3H_2O$$

Questions

1. What are the specific reactions involved in producing the color in each of the tests?

2. What difficulties might be encountered in quantifying the amount of protein using one of these qualitative color reactions?

■ EXPERIMENT 3: QUANTITATIVE DETERMINATION OF PROTEIN IN FOODS BY THE BIURET METHOD

Introduction and Objective

Substances containing two or more peptide bonds form a purple complex with copper (II) in alkaline solution. Color development is due to the coordination complex formed between the copper ion and nitrogen atoms in the peptide chain. The concentration of protein in a sample is obtained by reference to a calibration curve prepared with known concentrations of protein. The objective of this exercise is to quantitate the amount of protein in a food.

Reagents

- Bovine serum albumin or egg albumin standard: Dissolve 100 mg protein in 10 ml distilled water.

- Biuret solution: Dissolve 1.50 g $CuSO_4 \cdot 5H_2O$, 6.0 g of $NaKC_4H_4O_6 \cdot 4H_2O$ in 500 ml distilled water. Add 300 ml 10% NaOH with constant swirling. Dilute to 1 L with distilled water and store in a polyethylene bottle.

Completion time: 1 hour
Complications: none

Procedure

1. Pipette a known amount of protein standard and distilled water as shown in Table 13.1 into a test tube.

2. Add 4.0 ml of Biuret solution, mix, and allow to stand 30 min at room temperature.

3. Record optical density at 550 nm on a spectrophotometer (see Table 13.1).

4. Plot absorbance at 550 nm against the concentration of the protein standard.

TABLE 13.1

Concentration and Optical Density of Egg Protein Dispersions

Tube	Volume Standard (ml)	Volume Water (ml)	Concentration Standard (mg/ml)	Optical Density 550
1[a]	0.00	1.00	0.0	Set at 0
2	0.10	0.90	1.0	
3	0.25	0.75	2.5	
4	0.50	0.50	5.0	
5	0.75	0.25	7.5	
6	1.00	0.00	10.0	

[a] Reagent blank

5. Prepare an unknown that is neither heavily pigmented nor turbid, such as egg white, unflavored gelatin, pure protein solutions, or a semipurified protein dispersion such as a dialyzate. Estimate the appropriate dilution from a food composition table so that the final protein concentration will fall within the concentrations used for preparation of the standard curve. Pipette an aliquot into a test tube and add water if necessary to bring the total volume to 1.00 ml. Continue with Step 2 as for the standards.

■ EXPERIMENT 4: EFFECT OF HEAT ON PROTEINS

Introduction and Objective

Heat is one of the common denaturing agents of proteins in foods. Most proteins denature (uncoil) at a specific temperature. Once denatured, they are susceptible to coagulation/gelation, and this can be detected as turbidity in the dispersion. The objective of this exercise is to determine the effect of heat on the denaturation of egg albumin in aqueous solution.

Materials

- Egg — 1

- Water baths (set at 55, 60, 63, 65, and 68°C)

- Spectrophotometer

Completion time: 1.5 hours
Complications: none

Procedure

Prepare 100 ml of a 10% dispersion (v/v) of egg white and distilled water. Filter to remove opaque membranes. Place 5 ml of the albumin solution in each of five test tubes. Place tubes in a water bath. Save the remainder for an unheated control. Place one test tube at each of the following temperatures: 55, 60, 63, 65, and 68°C. Heat samples for 30 minutes. Cool the samples in tap water. Determine the absorbance of the samples with the spectrophotometer at 450 nm. Use the unheated sample to zero the spectrophotometer. Be sure to mix samples prior to readings.

Questions

1. Describe the appearance of the solutions.

2. Plot temperature vs. absorbance and compare the curve to visual observations.

3. From these data, what is the effect of heat on proteins?

■ EXPERIMENT 5: COAGULATION OF PROTEINS

Introduction and Objective

While heat is a common coagulating agent in food, other food components may also affect the extent to which proteins denature. Among these components are salts, sugars, and acids. The objective of this exercise is to investigate some of the factors that affect the coagulation of proteins.

Materials

- Egg white — 1

- Distilled water

- 0.1 M sodium chloride solution — 5 ml

- 0.1 M calcium chloride solution — 5 ml

- 0.1 M ferric chloride solution — 5 ml

- 0.1 M sucrose solution — 5 ml

- 1.0 M sucrose solution — 5 ml

- 0.001 M hydrochloric acid solution — 5 ml

- 0.1 M hydrochloric acid solution — 5 ml

Completion time: 1 hour
Complications: none

Procedure

Dilute one egg white (slightly beaten) with three volumes distilled water; stir slowly but thoroughly and filter. To each of a series of test tubes, add 10 ml of the albumin dispersion and 5 ml of each of the solutions listed above, including distilled water. Record the pH of the solutions containing distilled water and 0.01 M and 0.1 M HCl. Place all of the test tubes in a beaker of water, heat slowly, and note the temperature at which opalescence (cloudiness) develops. Generally, samples will need to be heated for at least 20 minutes.

■ EXPERIMENT 6: EFFECT OF PH ON HYDRATION OF MEAT PROTEINS

Introduction and Objective

The ability of muscle to retain water is an important property affecting the usefulness of muscle as a food. The extent to which muscle can retain water depends on a number of factors, among them pH. The objective of this exercise is to observe the effect of pH on hydration of the myofibrillar proteins, i.e., the water-binding capacity of meat.

Materials

- Ground beef, 25 g

- HCl, 6 N — 500 ml

- NaOH, 2 N — 500 ml

- Plastic centrifuge tubes, 6

- pH meter

- Centrifuge

Completion time: 1.5 hours

Complications: As this experiment requires a lot of weighing and pH measuring and adjusting, it should be started early in the lab period. Also, be sure to record the empty weight of each centrifuge tube.

Procedure

1. Label six 50-ml plastic centrifuge tubes with tape and then weigh each tube.

2. Transfer exactly 3 g of finely ground meat to each centrifuge tube. Add 15 ml of distilled water to each centrifuge tube and homogenize briefly with a Sorvall Omni-Mixer.

3. Adjust the pH of each tube to the following values using NaOH or HCl.

   Tube 1 — pH 4.0

   Tube 2 — pH 4.5

   Tube 3 — pH 5.0

   Tube 4 — pH 5.5

   Tube 5 — pH 6.0

   Tube 6 — pH 7.0

4. Check the pH of each tube 10 min after the initial adjustment.

5. After an additional 5 min, centrifuge the tubes for 10 min at 2000 r/min (approximately 1500× gravity). After centrifugation, carefully decant the supernatant and discard it. Accurately weigh the sedimented myofibrils.

Questions

1. What effect does pH have on the weight of the sedimented muscle?

2. How does this relate to hydration of the muscle proteins?

3. Explain and illustrate the chemical changes responsible for these effects. From your data, can you locate the isoelectric point of myofibrillar proteins?

■ EXPERIMENT 7: SPUN FIBER PRODUCTION

Introduction and Objective

Proteins, like other polymers, can be spun into fibers, e.g., in the production of textured vegetable protein (TVP), which is widely used in many types of food. The purpose of this exercise is to demonstrate the industrial process used in the manufacture of spun isolate fibers.

Materials

-   Precipitation solution (saturated aqueous NaCl solution adjusted to pH 2.5 with phosphoric acid)

-   Hypodermic syringe and #20 needle

-   10% NaOH

-   Soybean protein isolate (commercial preparation) — 3 g

Completion time: 30 minutes
Complications: none

Procedure

1. Suspend 3 g of protein isolate in 10% NaOH in a 250-ml beaker, adding NaOH until a soupy paste is formed.

2. Load the paste into a syringe and force the mixture slowly through the needle into the precipitation solution.

3. Repeat Step 2, but extrude the protein mixture into water.

Questions

1. Describe the appearance of the protein at each step of the spun fiber process.

2. Compare the strands to the material that was not exposed to the precipitation bath.

3. Why do proteins precipitate at their isoelectric point?

■ EXPERIMENT 8: EFFECTS OF THE ENZYME RENNIN ON MILK PROTEIN

Introduction and Objective

Rennin is an enzyme taken from a calf's stomach that is used in the production of cheese. Today, an enzyme made by biotech means called chymosin, which is identical to rennin, is widely used. The action of rennin/chymosin is highly dependent on temperature. The objective of this experiment is to investigate the effect of temperature on rennin activity.

Materials

- Whole milk — 350 ml
- Rennin tablets — 1

Variations

1. Use refrigerator temperature (5°C) milk and store rennin-treated milk in refrigerator.

2. Warm milk to 36°C before adding rennin.

3. Heat milk to boiling (100°C) before adding rennin.

Completion time: 1 hour and 30 minutes
Complications: none

Procedure

Bring milk (115 ml) to the desired temperature for your variation. Dissolve $1/4$ rennin tablet in 5 ml of water. When the milk reaches the desired temperature, add the rennin dissolved in the water. For the 5°C milk, store the milk–rennin mixture in the refrigerator. The other samples should be allowed to stand at room temperature for 1 hour. Observe each milk–rennin mixture at the end of one hour and describe its appearance (gelled, liquid, etc.). If any of the milk–rennin mixtures have gelled, measure the gel strength with the Texture Analyzer (record the results in Table 13.2). Refer to the Equipment Guide section for instructions for using the Texture Analyzer.

TABLE 13.2

Visual Appearance and Gel Strength of Milk Treated at Different Temperatures with Rennin

Milk Treatment	Visual Appearance	Gel Strength
5°C milk		
36°C milk		
100°C milk		

14

LABORATORY: GELATIN

Gelatin is a protein obtained from collagen (a type of connective tissue) that is used as a gelling agent. The viscosity of a protein sol such as gelatin varies with such factors as molecular size, molecular shape, temperature, degree of hydration, concentration, and pH. In general, as the hydrogen ion or hydroxyl ion concentration departs from the isoelectric point (IEP), absorption of water increases until a maximum is reached. Further increases result in hydrolysis of the gelatin. The method of preparation of the gelatin influences its isoelectric point. Acid-prepared gelatins have an IEP of 7 to 9, while alkaline-prepared gelatins have an IEP of 4.7 to 5.0. When a gelatin sol is cooled to below 35°C, the viscosity increases until gelation occurs. The effectiveness of gelatin as a gelling agent stems from its unique amino acid makeup: approximately one-third glycine or alanine, nearly one-fourth basic or acidic amino acids, and approximately one-fourth proline or hydroxyproline. Rigidity of the gel increases with greater concentration of the gelatin and is influenced by pH and the presence of other compounds such as nonelectrolytes (sucrose) or proteolytic enzymes.

EXPERIMENT 1: EFFECTS OF VARIATIONS IN GELATIN CONCENTRATION, pH, SUCROSE CONCENTRATION, AND PRESENCE OF A PROTEOLYTIC ENZYME ON GELATIN GEL STRENGTH

Introduction and Objective

Gelatin can gel in plain water. However, most food applications do not involve a system this simple. Typically, a complex food system will include, in addition to water and gelatin, such things as acids, sugars, and enzymes. These other components may well have an effect on gel setting and gel strength. The objective of this exercise is to determine the influence of gelatin concentration, pH, sucrose, and the presence of proteolytic enzymes on viscosity and setting time of a gelatin sol and to determine liquefying time and gel strength of a gelatin gel.

Materials

- Gelatin — 200 g
- 6 *M* HCl
- 2 *M* NaOH
- Sucrose — 70 g
- Papain, bromelain — 0.25 g
- Stopwatch
- 2-ml or 5-ml pipette

- Test tubes of equal diameter and rack

- Custard cups

Completion time: 2 hours

Complications: This experiment is best broken up among several groups of students due to the number of samples that are prepared and the number of observations that are taken.

Variations

1. Effect of concentration: Prepare a 6% gelatin solution by first dispersing 120 g gelatin in cold deionized water and then adding enough boiling deionized water to make 2000 g of sol. From this, prepare a series of dilutions of 500 ml each containing 6, 3, 1.5, and 0.5% gelatin.

2. Effect of pH: Prepare a gelatin solution by dispersing 37.5 g gelatin in cold deionized water. Add enough boiling deionized water to make 2000 g of sol. Divide into five equal portions. Adjust the five samples to pH 1, 5, 6, 7, and 12, respectively, using HCl or NaOH. Dilute each portion to 500 ml while still warm. Final concentration will be 1.5%.

3. Effect of sucrose: Prepare four 1.5% gelatin dispersions (500 g each) that vary in sugar concentration (0, 0.05, 0.1, and 0.2 M sucrose) by dispersing 7.5 g gelatin and sucrose (0, 8.6, 17.1, and 34.2 g) in cold deionized water. Add enough boiling deionized water to each preparation to make 500 g of sol.

4. Effect of proteolytic enzymes: Prepare a 1.5% gelatin solution by dispersing 7.5 g gelatin in cold deionized water. Add boiling deionized water to make 500 g of sol. Cool a little and then stir in 0.25 g of a proteolytic enzyme.

General Procedure

1. Once the test sols are prepared, pour sols while still warm into a custard cup, fill to exactly 1 cm from the top, and place in ice bath or refrigerator (as long as each variation is aged in the same way) to chill. Determine gel firmness with the penetrometer or the cone probe on the Texture Analyzer. Observe gels for clarity and firmness.

2. Pour 10 ml of each sol into a test tube and place the test tubes in a rack in a water bath at 10°C. Determine the setting time as the time elapses from 60°C until the contents of each tube cease to flow. When a gel has formed, invert the test tubes in a test tube rack placed over a paper towel at room temperature and determine liquefying time (when contents of tube reach the paper towel).

3. With excess sol, determine the time (use a stopwatch) of outflow of each sol in a 2- or 5-ml pipette (use the same size pipette for all sols) at 60°C. Make replicate readings.

4. Plot or graph the following:

   a. Concentration of gelatin vs. outflow time for sol

   b. Concentration of gelatin vs. setting time

   c. Concentration of gelatin vs. liquefying time

   d. Concentration of gelatin vs. gel strength

   e. pH vs. outflow time for sol

   f. pH vs. setting time

    g. pH vs. liquefying time

    h. pH vs. gel strength

    i. Concentration of sucrose vs. outflow time for sol

    j. Concentration of sucrose vs. setting time

    k. Concentration of sucrose vs. liquefying time

    l. Concentration of sucrose vs. gel strength

■ EXPERIMENT 2: EFFECT OF *IN SITU* ENZYMES ON GELATIN GEL STRENGTH

Introduction and Objectives

Gelatin gels may be affected by many variables, notably the presence or absence of proteolytic enzymes in materials added to the gelatin. The effectiveness of these proteolytic enzymes depends heavily on the previous thermal treatment of the food material containing the enzyme. The objective of this exercise is to investigate the effect of previous heating on the proteolytic enzyme in pineapple and, hence, the effect of added pineapple on gelatin texture.

Materials for Control Gelatin

- Unflavored gelatin — 1 g
- Room temperature water — 7 ml
- Boiling water — 15 ml
- Sucrose — 12.5 g
- Frozen orange juice concentrate (thawed) — 7 ml
- Ice water — 23 ml
- Lemon juice — 1 ml
- Fresh or frozen pineapple, pureed — 20 g
- Canned pineapple, pureed — 20 g

Variations

1. Control
2. Add fresh or frozen pineapple puree to beaker.
3. Add canned pineapple puree to beaker.

Procedure

1. In a 100-ml beaker, soak gelatin in room temperature water for 5 minutes.
2. Add the boiling water to the gelatin–water dispersion and stir until the gelatin disperses.
3. Blend in the sucrose, orange juice concentrate, ice water, and lemon juice.

4. To the control beaker add nothing. To one of the other two beakers add 20 grams of fresh or frozen pineapple puree and to the other beaker add 20 grams of canned pineapple puree. Stir well.

5. Set beakers at room temperature for 1 hour.

6. Record your visual observations on the gels and measure the gel strength using the Texture Analyzer (See Table 14.1). Refer to the Equipment Guide section for instructions on the use of the Texture Analyzer.

TABLE 14.1

Visual and Texture Observations on Gelatin Gels

Variant	Visual Observation	Gel Strength
Control		
Added fresh or frozen pineapple		
Added canned pineapple		

Questions

1. What is the relationship between the IEP of a protein and viscosity?

2. Describe the mechanism of action of each variable on a gelatin sol and gel.

3. What ingredients that would contain a proteolytic enzyme might be added to a gelatin product?

4. Suppose you were working for a company that manufactured gelatin desserts and you were asked to modify the product to decrease the time required for its preparation. What two possible changes could you suggest? Would these changes be desirable? Why or why not?

15 LABORATORY: CARBOHYDRATES

Carbohydrates, like proteins and fats, are major food constituents. They are divided into three main groups: monosaccharides, oligosaccharides, and polysaccharides. Monosaccharides are simple sugars that cannot be hydrolyzed into simpler compounds. Oligosaccharides yield two to ten simple sugars on hydrolysis. The reducing ability of a sugar depends on the presence of a potentially free carbonyl group in its structure. Reducing sugars participate in Maillard browning. Polysaccharides yield a large number of sugars upon hydrolysis. Starches, with their ability to bind large amounts of water, are useful polysaccharide thickening agents. Commercial starches are obtained from cereal grains, root sources, or trees. Each starch has its own characteristics.

■ EXPERIMENT 1: FEHLING'S TEST FOR REDUCING SUGARS

Introduction and Objective

The reducing ability of sugars can be determined by their ability to reduce cupric ions in alkaline solution to form red cuprous oxide in the reaction:

$$2\ Cu^{2+} + \text{reducing sugar} \longrightarrow Cu_2O \downarrow + \text{oxidized sugar}$$

Blue solution   Red precipitate

    The best-known reagent for this test is Fehling's solution. A positive reaction is shown by a green, yellowish-orange, or red coloration of precipitate, depending upon the amount and kind of reducing sugars. The degree of color change from blue to red is an indication of the extent of the reduction and therefore a measure of the reducing power of the sugar. The objective of this exercise is to test a variety of carbohydrates to determine whether or not they are reducing sugars.

Materials

- A variety of sugars (mono- and disaccharides) and sugar alcohols — 50 mg of each
- Crystalline copper sulfate — 35 g
- Sodium potassium tartrate — 180 g
- Sodium hydroxide — 50 g

Completion time: 30 minutes
Complications: none

Procedure

1. Dissolve 34.64 g of crystalline copper sulfate in 500 ml of water and label as solution "A."

2. Dissolve 173 g of sodium potassium tartrate and 50 g of sodium hydroxide in 500 ml of water and label as solution "B."

3. Mix 2.5 ml of solution A with 2.5 ml of solution B in a test tube. Add 10 to 50 mg of each sugar and boil for 5 to 10 min.

Questions

1. What is a reducing sugar? What is the significance of the reducing ability of a sugar?

2. Draw the structure of each sugar tested and indicate the reducing potential of each sugar. Were your results in agreement with those expected?

■ EXPERIMENT 2: MICROSCOPIC APPEARANCE OF STARCH

Introduction and Objective

Most, if not all, of the interesting applications of starches in foods depend on their being cooked. Sometimes this cooking occurs in the presence of acids or starch-degrading enzymes (amylases). When these treatments happen, the appearance of the starch granules changes. The objective of this exercise is to visually assess physical changes in starches exposed to various heating and/or chemical treatments.

Reagent

- Prepare a 2% solution of KI and add sufficient iodine to color it a deep yellow.

Completion time: 1 hour
Complications: none

Procedure

Under the microscope, examine slides of cornstarch and waxy cornstarch stained with iodine. A few granules of the uncooked starches can be placed on the slides or a thin smear of the starch pastes prepared in various steps of Experiment 3 may be used. Compare various uncooked starches and cornstarch exposed to the following treatments:

1. Uncooked

2. Paste heated to pasting temperatures (Get samples from Experiment 3, variations a and b.)

3. Paste heated to pasting temperature in the presence of acid (Experiment 3, variation l)

4. Paste heated to pasting temperature in the presence of diastase (Experiment 3, variation n)

Observe and sketch the color, shape, and size of the starch granules. Unbranched starch molecules (amylose) give a blue color, and branched starch molecules (amylopectin) give a reddish-black color.

■ EXPERIMENT 3: STARCH GELS

Introduction and Objective

The principal tasks that starches are asked to perform are thickening and gelling. Starch can thicken or gel water by itself, but real food systems are always more complicated than this. The objective of this exercise is to investigate the factors that affect the thickness of a cooked starch paste and to compare the usefulness of various starches (normal, waxy, and modified) as thickening agents (variations noted in Table 15.1).

Formulas

See Table 15.1.

TABLE 15.1

Starch and Other Ingredient Variations

	Type of Starch	Endpoint Temperature	Weight of Starch (g)	Added Ingredients
a.	Cornstarch	95°C	36	530 ml H_2O
b.	Waxy cornstarch	75°C	36	530 ml H_2O
c.	Wheat	95°C	36	530 ml H_2O
d.	Waxy cornstarch, slightly cross-bonded (Cerestar StabiTex 06330)	78°C	36	530 ml H_2O
e.	Waxy cornstarch, modified (Cerestar PolarTex 06732)	75°C	36	530 ml H_2O
f.	Potato starch	70°C	36	530 ml H_2O
g.	Tapioca starch	85°C	36	530 ml H_2O
h.	Cross-linked tapioca (Cerestar CreamTex 75710)	85°C	36	530 ml H_2O
i.	Cornstarch	95°C	36	530 ml H_2O, 75 g sucrose
j.	Cornstarch	95°C	36	530 ml H_2O, 53 g fat
k.	Cornstarch	95°C	36	530 ml H_2O, 3.18 g glyceryl monostearate
l.	Cornstarch	95°C	36	530 ml 0.5 N citric acid
m.	Cross-linked tapioca (Cerestar CreamTex 75710)	85°C	36	530 ml 0.5 N citric acid
n.	Cornstarch	95°C	36	530 ml 0.25% diastase solution

Completion time: 2 hours

Complications: This experiment is best broken up among several groups of students due to the number of samples that are prepared and the number of observations that are made. Also, be sure to label starch pastes to be stored (Step 4) with sample name, date, and group number.

Procedure

1. Disperse granules in cold liquid in heavy aluminum pan.

2. Cook over direct heat (medium low), stirring two turns per 15 seconds for the first minute and two turns per minute until endpoint temperature is reached. Observe hot paste for clarity. Those preparing variations a, b, l, and n also need to remove samples for microscopic observation (Experiment 2) at their respective pasting temperatures. Observe hot pastes for clarity.

3. Pour each hot starch paste into a Pyrex custard cup, fill to top, and level as carefully and accurately as possible. Place in refrigerator or ice bath and chill to 20°C. Note and record clarity of gel and then measure firmness with a penetrometer or the cone attachment of the Texture Analyzer and determine percent sag with a Vernier caliper. Consult the Equipment Guide section under penetrometer or Texture Analyzer and Vernier caliper for use of these instruments.

4. For each variation, fill two labeled 2-ounce plastic soufflé cups with hot starch paste, cover tightly with close-fitting lids to prevent evaporation; refrigerate one sample and freeze the other for 1 or more days. Thaw frozen samples at room temperature and examine the gel strength and sponginess of each sample.

5. Cool excess paste to 60°C and determine viscosity by linespread (allow paste to spread exactly 2 min).

Questions

1. Which starches would be suitable for thickening a white sauce or gravy? For a cherry pie? For a cream pie? For a sauce to be frozen? Explain your answers.

2. How well did objective data correlate to observations of the effect of refrigeration and freezing on the starch pastes?

3. A starch gel with and without acid showed the objective data listed in Table 15.2. Do the three readings in the table correlate with each other? Explain. Could you have drawn similar conclusions from your results?

TABLE 15.2

Linespread and Penetrometer Data for Starches Cooked With and Without Acid

	Linespread (mm)	Penetrometer (1/10 mm)	Sag (%)
With acid	5	80	30
Without acid	4	60	20

4. What is the effect of cross-linking on resistance of a starch paste to addition of acid?

■ EXPERIMENT 4: VISCOSITY CURVES OF STARCH PASTES

Introduction and Objective

Starch dispersions undergo a series of physical and structural changes upon gelatinization. Accompanying these changes are a series of changes in dispersion viscosity. The Brabender amylograph is an excellent instrument with which to observe these viscosity changes as a function of temperature and time. The objective of this exercise is to observe changes in apparent viscosity during heating, holding, and cooling of starch pastes.

Materials

- A variety of starches (cornstarch, waxy cornstarch, potato starch) — 35 grams of each
- Brabender amylograph or VISCO/amylo/GRAPH

Completion time: 1.5 hours
Complications: none

Procedure

1. Add one of the starches listed in Table 15.3 to 450 ml water. Stir and add to the amylograph bowl. For use of the Brabender VISCO/amylo/GRAPH, refer to the Equipment Guide section.

TABLE 15.3

Starches to Be Used in the VISCO/amylo/GRAPH Experiments

Type of Starch	Quantity (g)
Cornstarch	35
Waxy cornstarch	35
Potato	20
Tapioca	35
Wheat	35
Cerestar StabiTex 06330	35
Cerestar CreamTex 75710	35

2. Begin heating at room temperature or 25°C and continue heating at 1.5°C per minute until the gelatinization peak is reached or until 95°C.

3. Maintain this temperature for 15 to 20 min while stirring and recording the viscosity continuously.

4. Cool the paste to 50°C at a rate of 1.5°C per minute.

5. Label the Brabender curves with the following information:

   a. Peak viscosity

   b. Breakdown

   c. Setback

Questions

1. Compare the starches in terms of gelatinization temperature, rate of gelatinization, peak viscosity, breakdown, setback, and stability of the cooked paste.

2. How do cereal starches and root starches differ in terms of gelatinization temperatures?

3. What does the breakdown of a paste indicate? Why is setback of a paste of interest? When would a starch with a high setback be desirable?

4. Correlate the Texture Analyzer/penetrometer and linespread readings on the cooked pastes from Experiment 3 with the final viscosity of the cooled pastes. Can these values be used to determine whether or not a starch is a good choice for use in a cream pie or fruit pie?

16 LABORATORY: FLOUR MIXTURES

Wheat flour is derived from the cereal wheat, which is commonly used in the United States. For baked products to have a satisfactory texture, it is necessary to have a structural network that is flexible and rather stretchy during baking yet is reasonably strong after the product is removed from the oven. Gluten, developed from flour, is primarily responsible for the elastic character of batters and doughs during baking and for their semirigid structure after baking. Gluten development is influenced by the presence of added ingredients.

■ EXPERIMENT 1: GLUTEN BALLS

Introduction and Objective

When wheat flour is mixed with water, the proteins glutenin and gliadin react to form gluten, which is responsible for providing structure to baked goods as well as trapping leavening gasses. However, most batters and doughs are more complex than just wheat flour and water. The other ingredients, such as sugar, oil, salt, emulsifiers, and dough conditioners, influence the amount of gluten that is developed in the stirring/kneading process and thus the final texture of the baked product. The objective of this exercise is to demonstrate the effect of added ingredients on gluten formation.

Materials

- Whole wheat flour — 50 g
- Bread flour — 50 g
- All-purpose flour — 50 g
- Cheesecloth
- KI–Iodine solution — 5 ml
- Sugar — 25 g
- Vegetable oil — 10 ml
- Salt — 1 g
- Monoglycerides — 100 mg
- Sodium stearoyl lactylate (SSL) — 100 mg

Completion time: 1.5 hours
Complications: none

Formulas (See Table 16.1)

TABLE 16.1

Formulas to Be Used in Preparation of Gluten Balls

Variation	Flour (50 g)	Added Ingredient	Water (ml)
1	Whole wheat	—	30
2	Bread flour	—	30
3	All-purpose	—	30
4	All-purpose	25 g sugar	30
5	All-purpose	10 ml oil	20
6	All-purpose	0.025 g monoglycerides	30
7	All-purpose	0.025 g SSL	30
8	All-purpose	0.5 g NaCl	30

Procedure

1. Mix flour and added ingredients thoroughly. Stir in water to produce a dough that can be kneaded with the hands.

2. Knead the dough for 10 to 15 min until gluten is well developed.

3. Put the dough in a bag made of a double thickness of cheesecloth and wash the bag of dough under running water until the wash water remains completely clear. Check clarity by washing in a beaker of water until no purple color appears when the wash water is treated with I_2-KI. This step may take a half hour or more. **Wash the bran out of the whole wheat dough before placing it in the cheesecloth.**

4. When washing is complete, weigh the material in the dough bag. This is gluten. Notice the elasticity of the product.

5. Form the dough into a ball and bake at 450°F for 15 min; lower oven temperature to 300°F and bake 20 min longer or until dry. The oven should not be opened during baking. Therefore, all gluten balls to be placed in the same oven should be placed on a cookie sheet at least 6" apart and put in the oven *at the same time*. Observe the relative sizes of the baked gluten balls.

6. Prepare a bar graph of the weight of baked gluten vs. the variation.

Questions

1. What is the effect of each variation on gluten development?

2. What substance is in the washings that imparts the blue color with iodine? What process would be observed if the washings were boiled?

■ EXPERIMENT 2: SUGAR COOKIES

Introduction and Objective

Cookies can be used as a model system to show the effects of variations in recipe ingredients. The recipe can be varied by using higher protein all-purpose flour instead of the lower protein cookie flour, using extra fat, varying baking times and temperatures, varying prebake cookie thickness, and making the cookies with sugar and/or fat replacers. All these variations will make a difference in the texture and/or taste of the final product. The objective of this exercise is to illustrate the effect of variations in major ingredients and methodology on the quality and caloric value of sugar cookies.

Materials for Basic Recipe

- 55 g cookie flour
- 24 g hydrogenated fat
- 33 g sugar
- 8 ml milk
- 12 g blended whole egg
- 0.75 g salt
- 0.25 g baking soda
- 1 g S.A.S. baking powder
- 0.6 ml vanilla
- 0.25 g cinnamon
- 0.25 g nutmeg

Completion time: 2 hours

Complications: This experiment has a lot of variations and is best conducted by dividing it up among several groups of students in the lab.

Variations

1. Basic recipe.

2. Use 55 g of all-purpose flour in place of cookie flour.

3. Use 70 g cookie flour.

4. Use 48 g of fat.

5. Make the basic recipe, but bake at 350°F for 10 to 12 min.

6. Make the basic recipe, but in Step 4 of the Procedure place two pastry slats on each side and roll to that thickness.

7. 50% fat replacement: Substitute 24 g hydrogenated fat with the following — 12 g hydrogenated fat + 12 g 25% w/v Oatrim gel. **(Prepare the day before by adding Oatrim powder to hot water slowly while blending in a blender.)**

8. 50% fat and 50% sugar replacement: Substitute 24 g hydrogenated fat and 33 g sugar with the following:

   - 12 g hydrogenated fat +
   - 12 g of 25% w/v Oatrim gel +
   - 16.5 g sugar +
   - 1.5 g Sweet One

9. 100% sugar substitution: substitute 33 g of sugar with the following:

   - 29.9 g Litesse + 3.1 g Sweet One

Procedure

1. Sift the flour, salt, soda, baking powder, and spices together on paper.

2. Mix the fat, vanilla, and sugar together in a bowl. Add the egg and milk. Stir well.

3. Add the sifted dry ingredients from Step 1, and stir until the ingredients are well blended.

4. Weigh 20-g portions of the dough, form into balls, and roll to a uniform height with a rolling pin using pastry slats as a guide. (*Note:* students doing variation 6 should follow the specific rolling instructions for that variation.) Place onto cookie sheets 2" apart.

5. Bake at 400°F for 6 to 8 min. (*Note:* students doing variation 4 should use the baking time and temperature indicated for that variation.)

6. Measure breaking strength of at least two cookies from each variation with a shortometer or the knife probe of the Texture Analyzer. Refer to the Equipment Guide section for use of shortometer or Texture Analyzer.

Questions

1. Which variations affected texture? Color? Flavor? Explain how the variations effected a change in these attributes.

2. What is the chemical nature of the sugar and fat substitutes?

■ EXPERIMENT 3: CHOCOLATE CAKES

Introduction and Objective

Cakes are also prime candidates for ingredient variation. Higher-protein flour variants can be used in lesser amounts, leavening (baking soda) can be altered, or monoglycerides can be added. The objective of this exercise is to determine the effect of altering ingredients and the effect of pH variation on the color, flavor, and texture of chocolate cake.

Materials for Basic Recipe

- 73 g shortening
- 175 g sugar
- 2 g salt
- 2 ml vanilla
- 64 g eggs
- 1 oz baker's chocolate
- 39 ml boiling water
- 79 ml milk
- 2 g baking soda
- 112 g cake flour

Completion time: 2 hours
Complications: none

Variations

1. Basic recipe.

2. Substitute all-purpose flour for cake flour.

3. Use 1 g soda.

4. Use 4 g soda.

5. Add 0.056 g monoglycerides.

Procedure

1. Prepare pan by oiling and fitting bottom with waxed paper.

2. For variation 5, mix monoglycerides with shortening. Cream shortening, eggs, sugar, vanilla, and salt for 15 min (medium speed electric mixer).

3. Pour boiling water over chocolate, stir until smooth. Let cool to 50°C; stir cooled chocolate into creamed mixture on medium speed for 1 min.

4. Sift the baking soda twice with the flour.

5. Add flour and milk alternately (in thirds) to the creamed mixture, stirring a total of 1 min following each addition and a total of 2 min following the last addition using medium speed on the hand or bowl mixer.

6. Bake two 150-g loaf cakes for each variation. Bake cakes at 365°F for approximately 18 to 20 min. Test with toothpick for doneness or use thermocouple with recording potentiometer.

7. Use remaining batter for pH readings and to determine specific gravity of the batter. Refer to Equipment Guide section for use of pH meter and to determine specific gravity of solids.

8. Determine volume of cake using the seed volume apparatus (see Equipment Guide).

9. Cut 1-in. slices from a 150-g loaf using the cutting block, and determine the area of a slice as an index to volume with the compensating polar planimeter. To assess crumb firmness, cut three cylinders from a 1-in. slice and determine compressibility using the compressometer with the flat disc attachment or the Texture Analyzer with the cylinder attachment. Refer to the Equipment Guide section for use of the compensating polar planimeter and compressometer or Texture Analyzer.

10. Photocopies may be used to compare cell structure.

11. Rank each cake for uniformity and size of cells, moistness of crumb, compactness, darkness of color, and flavor.

12. Cut slices of each cake that will cover the bottom of a Petri dish and take L, a, b color readings using the Hunter Colorimeter. Refer to the Equipment Guide section for use of the Hunter Colorimeter.

Questions

1. What is the effect of pH on chocolate flavor and color?

2. Does this formula have a high or low sugar-to-flour ratio? Would this make a difference in the effect of ingredient or procedural manipulation on the outcome of the product?

17 LABORATORY: PIGMENTS

Pigments contribute greatly to the aesthetic appeal of foods. The chemical forms of some pigments are easily altered under conditions that may also affect the structural integrity of the tissue. Heating, pH changes, and oxidation reactions can affect pigment quality. The predominant meat pigment is myoglobin. Reactions of myoglobin determine the color of fresh and cured meats. Plant pigments may be categorized as carotenoids, chlorophylls, and flavonoids. Included in the flavonoid group are the phenolic compounds, which are the substrates in the enzymatic browning of fruits and vegetables. Preservation of desirable color, flavor, and textural qualities present at harvest of ripe fruits and vegetables depends greatly on control of the deteriorative changes caused by endogenous enzymes. Sometimes colorants are added to foods to enhance their marketability.

■ EXPERIMENT 1: COLOR REACTIONS OF MYOGLOBIN

Introduction and Objective

An important discoloration reaction in fresh meat is oxidation of reduced myoglobin to metmyoglobin, a brown pigment. This reaction also occurs during color fixation in curing meats, since the initial action of sodium nitrate on myoglobin is to oxidize the iron to the ferric form. Ferric iron must be reduced to the ferrous form to permit formation of the cured meat pigment, nitric oxide myoglobin. Reduction of the ferric iron in this experiment is accomplished by the sulfhydryl groups that become available upon heat denaturation of protein or from added chemical reducing agents such as ascorbic acid. If the concentration of sodium nitrite is too high or if ascorbic acid is present in the absence of sufficient sodium nitrite, undesirable reactions can occur with the production of green pigments due to the opening of the porphyrin ring of myoglobin. The objective of this exercise is to demonstrate some of the reactions of a heme pigment.

Materials

- Beef — 25 g
- Sodium nitrite — 100 mg
- Sodium hydrosulfite or sodium dithionite —100 mg
- Ascorbic acid — 100 mg

Completion time: 1.5 hours
Complications: none

Extraction Procedure

1. Grind fresh meat.

2. Homogenize meat in 100 ml of cold water. Keep cold (4°C) while blending. Generally, 25 grams of ground beef in 100 ml of cold water will produce enough supernatant for all the tests required.

3. Centrifuge homogenate at 1000 × g for 10 minutes.

4. Decant supernatant and filter through glass wool.

Treatments

Prepare two 5-ml samples of each of the specific chemical treatments. One sample of each should be placed in a hot H_2O bath (75°C) for 1 hour, the other held at room temperature.

1. Color reactions in fresh meat:

   - 30 mg sodium hydrosulfite or sodium dithionite

   - Supernatant only

   - 15 mg $NaNO_2$

   - 50 mg $NaNO_2$

2. Color reactions in cured meat:

   - 30 mg ascorbic acid

   - Add 15 mg $NaNO_2$ and mix; then add 30 mg ascorbic acid

   In each case, note the initial color after addition of the reagents and any color changes following mixing and following the start of heating. Observe against a white light if necessary to see color changes. Record the water bath temperature.

Questions

1. What types of reactions are responsible for the colors observed in the various treatments? What is the name of each pigment formed? Which reactions are desirable and which represent common discoloration reactions? What is the probable oxidation state of the iron, and is the porphyrin ring intact?

2. What is the effect of heating on the globin portion of the pigment?

■ EXPERIMENT 2: THE EFFECTS OF HEAT AND pH ON PLANT PIGMENTS

Introduction and Objective

Many plant pigments, especially chlorophyll and the anthocyanins, are sensitive to heat and changes in pH. Under favorable acid conditions, these pigments will exhibit their correct colors, but when the pH is increased or decreased, the pigment may change to an undesirable color. This presents a sensory defect in the food. The objective of this exercise is to determine the effect of heat and pH on plant pigments.

Materials

- Frozen peas — 25 g
- Canned peas — 25 g
- Vinegar — 10 ml
- Grape juice — 10 ml
- Cranberry juice — 50 ml
- 1 N NaOH — 100 ml

Completion time: 1.5 hours
Complications: none

Procedure: Chlorophyll

1. Heat 150 ml deionized water to boiling.

2. Add approximately 25 g frozen peas.

3. When the water returns to a boil, time for 7 min.

4. Remove the sample from the water and place in a beaker.

5. Add 10 ml vinegar to about 150 ml deionized water and determine the pH of the solution. Boil the solution and repeat Steps 2 to 4.

6. Add 10 ml 1 N NaOH to 150 ml deionized water and determine the pH of the solution. Boil the solution and repeat Steps 2 to 4.

7. Expose a fourth 25-g sample of peas to a cold mixture of 10-ml vinegar and 150-ml deionized water for 7 min without cooking.

8. Expose a fifth 25-g sample of peas to a cold mixture of 2 g $NaHCO_3$ and 150 ml deionized water without cooking.

9. Set up a sixth beaker with canned peas.

10. Compare all the samples for color and texture.

Procedure: Anthocyanins

1. Mix 10 ml of grape juice and 90 ml of distilled water.

2. Determine the pH of the solution.

3. Remove 5 to 10 ml of this solution to a test tube.

4. Adjust the remaining solution to pH 5.0 with 1 N NaOH.

5. Remove 5 to 10 ml of this solution to a test tube.

6. Adjust the remaining solution to pH 7.0.

7. Remove 5 to 10 ml of the solution to a test tube.

8. Adjust the remaining solution to pH 10.0.

9. Remove 5 to 10 ml to a test tube.

10. Compare all the samples, especially noting the color.

11. Mix 50 ml cranberry juice and 50 ml of distilled water.

12. Repeat Steps 2 to 10 with cranberry juice.

Questions

1. How do vegetable textures and the color of chlorophyll change when plant tissue is subjected to heat at various pH values? What chemical changes occur in the chlorophyll molecule in the various treatments? What changes account for the textural changes?

2. Illustrate the chemical effects that changes in pH have on the structure of anthocyanin molecules and the color of anthocyanin solutions.

■ EXPERIMENT 3: SEPARATION OF PIGMENTS IN A GREEN, LEAFY VEGETABLE

Introduction and Objective

Pigments, like other organic compounds, can be separated by various chromatographic techniques. Chromatography takes advantage of the differential solubilities in the mobile phase and differential attractions to the solid phase to separate compounds. The solid phase can be a number of things including cellulose (paper) and silica gel. The objective of this exercise is to extract and separate by paper or thin layer chromatography (TLC) the main pigments from broccoli, spinach, or other green, leafy vegetables.

Materials

- Green, leafy vegetable — 3 g

- Petroleum ether — 250 ml

- Acetone or ethanol — 10 ml of either

- *n*-Butanol — 1 ml

- 6-in. test tubes with stoppers

- Transfer pipettes, capillary tubes, or micropipettes

- Whatman No. 1 filter paper

Completion time: 1.5 hours

Complications: Be careful when using the organic solvents in this experiment. Be sure to dispose of the organic solvents correctly when you are done with them.

Procedure: Extracting the Pigments

1. Cut a piece of green leaf approximately 2 in. square. Drop it into a small amount of boiling water and boil for 2 min.

2. Remove leaf from the water and pat it dry between pieces of paper toweling.

3. Combine 3 ml of petroleum ether and 10 ml of either acetone or ethanol in a test tube. With scissors, shred the leaf into the test tube. Make sure the pieces are covered with solvent.

4. Let stand a few minutes until most of the pigments are dissolved and the leaf has lost most of its color. Stir once or twice with a glass stirring rod.

5. Remove the liquid to another test tube with a transfer pipette. Add 10 ml of water, stopper, and invert the test tube several times. Let stand until the two phases separate.

6. With a transfer pipette, transfer the upper dark-green layer to a clean test tube. Add 10 ml of water, stopper, and invert the test tube several times. Again let stand until the two phases separate.

7. Repeat the washing, as in Step 6. The pigments are now in the petroleum ether phase and the water has removed most of the acetone or ethanol.

8. If the pigments do not separate on the chromatogram (see Step 8 in the chromatographic separation procedure below), repeat the washing procedure.

Procedure: Chromatographic Separation of the Pigments

1. Prepare the developing fluid. Pipette 1 ml of *n*-butanol into a 100-ml volumetric flask and bring to volume with petroleum ether. Shake well. *Caution:* Keep away from flame.

2. Pour aliquots of the developer into 6-in. test tubes to a depth of $\frac{1}{2}$ in. Stopper with corks.

3. Cut several strips 6 in. long by $\frac{1}{2}$ in. wide from Whatman No. 1 filter paper. Make a fold $\frac{1}{2}$ in. from one end of each strip, and with a pencil place dots in the center and $\frac{1}{2}$ in. from the other end of the strip.

4. With a micropipette or capillary tube, apply a small drop of the petroleum ether extract of the pigments to the filter paper strip at the position marked by the dot. Keep the spot small (3/16 in. in diameter, see Figure 17.1). Make four to eight applications at the same spot, but allow to dry for a few seconds between applications. Practice spotting with a scrap of filter paper first.

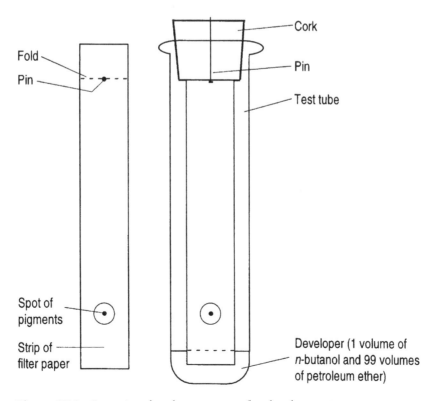

Figure 17.1 Preparing the chromatogram for development.

5. With a straight pin, attach the strip of filter paper at the center of the crease to the bottom of the cork. Make sure the strip hangs straight.

6. Lower the strip into one of the test tubes that contains developer. The developing fluid should reach the end of the filter paper strip, but it should not touch the spot of pigments.

7. Let the chromatogram develop until the pigments are separated.

8. Appearing on the chromatogram of extracts from most green leaves are carotenes, which move with the solvent front. Blue-green chlorophyll a moves ahead of the olive-green chlorophyll b. A faint gray band of pheophytin a may be observed in front of chlorophyll a. One or more spots or bands of yellow xanthophylls also may appear on the chromatogram.

9. Draw a diagram of the chromatogram and label the spots.

Thin Layer Chromatography

As an alternative to paper chromatography, the spinach pigments may be separated by thin layer chromatography. Spot a small aliquot of the petroleum ether extract of the pigments approximately 1.5 cm from the end of a silica gel chromatography plate. Keep the sample spot 1 cm or smaller in diameter.

When the sample spot is thoroughly dry, place the plate in a developing chamber containing solvent (petroleum ether–isopropanol–water, 100:5:0.25, v/v/v). Prepare the solvent by adding the isopropanol to the water and then the petroleum ether to this mixture. The depth of the solvent is critical since the sample spot must not dip into the solvent. When the solvent front has advanced 15 cm above the origin, remove the chromatogram from the developing tank and allow it to dry.

Identify as many of the pigments as possible.

■ EXPERIMENT 4: ENZYMATIC BROWNING

Introduction and Objective

Some plant tissues contain phenolics associated with their cell walls. Some of these also contain polyphenoloxidase (PPO), an enzyme that will convert the phenolic to a quinone, which will eventually be transformed into a brown melanoidin pigment. This reaction is generally undesirable when it occurs in tissues of fruits such as apple, banana, or pear. It is thus important to know how to control this browning reaction. The objective of this exercise is to assess the effect of various treatments on enzymatic discoloration of Red Delicious apples.

Materials

- 1 Red Delicious apple
- 1% thiourea — 60 ml
- Ascorbic acid — 10 mg
- Sodium sulfite — 10 mg
- Dipotassium phosphate — 120 mg
- Spectrophotometer
- Buchner funnel and filter flask
- Whatman No. 1 filter paper
- Blender

Completion time: 1.5 hours

Complications: none

Procedure

Note: It is important to work quickly once the apple tissue is cut until the slices are placed into solution. Have the beakers labeled and solutions prepared in advance.

1. Peel and pare apples, cut into uniform thin slices, and divide into four lots of 30.0 g each. Place one lot into a beaker containing 60 ml 1% thiourea, which will stop the browning reaction and serve as the control. Place the second lot into a beaker containing 60 ml deionized water. Place the third lot into a beaker containing 60 ml deionized water with 0.01 g ascorbic acid. Place the fourth lot into a beaker containing 0.01 g sodium sulfite in 60 ml water, and after 45 seconds decant the solution and replace with a solution containing 0.12 g dipotassium phosphate in 60 ml water.

2. After the apple tissue has been in solution for 30 min, homogenize the contents of each beaker in a blender, and filter through a Buchner funnel into a filtration flask under aspirator vacuum using Whatman No. 1 filter paper. Attach the funnel to the filtration flask, turn on the aspirator, wet a piece of filter paper using a squeeze bottle of deionized water, and carefully place the filter paper in the Buchner funnel. Pour apple tissue from the blender into the Buchner funnel and continue filtration until several milliliters of filtrate have been collected.

3. Place 1 ml of each filtrate in four different test tubes each containing 5 ml water and mix.

4. Transfer contents of each test tube to a cuvette and read optical density at 475 nm using a spectrophotometer. Use deionized water to set the instrument to 0% T. Refer to the Equipment Guide for use of the spectrophotometer.

Questions

1. Outline the reactions that occur as apples turn brown. Which intermediate is being detected by the spectrophotometer?

2. Describe the effect of each treatment on discoloration and show on the reaction diagram the point of action of each.

■ EXPERIMENT 5: PEROXIDASE ASSAY TO DETERMINE ADEQUACY OF BLANCHING

Introduction and Objective

Blanching is a heating process performed prior to freezing that inactivates enzymes that would otherwise cause deterioration in palatability, color, and ascorbic acid content during storage. Polyphenol oxidase (PPO), which catalyzes the enzymatic browning reaction studied in Experiment 4, is one target enzyme to inactivate during blanching. Catalase and peroxidase are two endogenous enzymes that are frequently used as indices of the adequacy of blanching treatment because they are more heat resistant than PPO. It is assumed that if these enzymes are inactivated, enzymes that catalyze undesirable color, flavor, and texture changes will also be inactivated. The objective of this exercise is to test for the presence of peroxidase in raw and blanched samples of a vegetable.

Materials

- Broccoli — 1 head

- Guiacol — 1 ml

- 0.08% H_2O_2 (2.7 ml of 3% H_2O_2 diluted to 100 ml with deionized water) — 10 ml

- Watch glasses — 2

- Blender

- Funnel

- Glass wool

Completion time: 30 minutes

Complications: Do the part of the experiment using guiacol in a fume hood.

Procedure

1. Preparation of samples

   a. Divide broccoli into two lots. Remove large leaves and lignified parts of stalks. Cut broccoli lengthwise into uniform pieces. Blanch one lot by immersing in boiling water for 3 min. Remove from boiling water and plunge into cold water until cool. Let drain.

   b. For each lot, blend one part vegetable with three parts distilled water (by weight) in a blender for 3 min. Filter through a funnel with a glass wool plug.

2. Assay for peroxidase

   a. Put several milliliters of filtrate from the raw broccoli on a watchglass.

   b. Add several drops of guiacol and several drops of 0.08% H_2O_2. A brick-red color indicates peroxidase activity.

   c. Repeat Steps 2a and b for the blanched broccoli.

Questions

1. Was the blanching treatment adequate?

2. Why are catalase and peroxidase termed index enzymes to determine adequacy of blanching?

▬ EXPERIMENT 6: MEASUREMENT OF COLOR OF ORANGES

Introduction and Objective

Sometimes oranges or grapefruit ripen in terms of flavor and texture, but their skins are not completely colored. Ethylene gas can be used to hasten development of the orange color, or artificial color can be used. When artificial color is used, the oranges are individually stamped "color added" or "artificially colored." Uncolored oranges or grapefruit may have green streaks even though they are ripe. The Federal Food, Drug and Cosmetic Act allows the addition of color on ripe and otherwise acceptable quality oranges or grapefruit. Color cannot be used to conceal inferiority or damage. Citrus Red No. 2 is often used. The amount used must be within safe tolerance limits. The objective of this exercise is to determine whether a red color has been added to stamped or unstamped oranges or grapefruit.

Materials

- Whatman No. 3 MM filter paper
- Citrus Red Dye No. 2 dissolved in chloroform — 5 ml
- Transfer pipettes
- Acetone — 130 ml
- 100-ml graduated cylinders
- 100-ml beakers
- Ethyl ether — 70 ml
- Chromatographic tank
- 125-ml Erlenmeyer flask
- Funnel
- Glass rods
- Chloroform
- Pipette
- Oranges and/or grapefruits, 3 stamped "color added" and 3 not stamped

Completion time: 1.5 hours
Complications: none

Procedure

1. Place a 7-in. × 9-in. sheet of Whatman No. 3 MM filter paper so that the 9-in. dimension is vertical. Use a soft lead pencil to draw a horizontal line across the sheet 1-in. from the bottom edge. Mark off three $1/2$-in. areas for spotting solutions. Label these areas: Citrus Red Dye No. 2, "color added" oranges (or grapefruit), and oranges (or grapefruit) not stamped "color added."

2. Place a 125-ml Erlenmeyer flask beneath a 125-ml filtering funnel and support an unstamped orange or grapefruit on three glass rods. Wash the color off by spraying the fruit with 25 ml of chloroform in the form of a fine stream from a pipette.

3. Repeat Step 2 with two more pieces of fruit, combining the washings in the same flask.

4. Concentrate the washings in a rotary evaporator, but do not let all of the solvent evaporate.

5. Repeat Steps 2 to 4 with oranges or grapefruit stamped "color added."

6. Pour a solution of 130 ml acetone and 70 ml ethyl ether into the bottom of a chromatographic tank. Spot approximately 50 μl of Citrus Red Dye No. 2 and the unknowns dissolved in a little chloroform on the marked sheet using capillary pipettes. Place the sheet in the chromatographic tank and suspend from a rod so the bottom edge of the paper is immersed about $1/4$" in the solvent. Cover the tank and allow the paper to remain undisturbed for about 1 hour.

7. Remove paper and air dry. Measure the distance the colors have traveled. By comparing the distance traveled, determine the dye added, if any. Natural coloring materials present will remain at the origin.

18 LABORATORY: PECTIN

Pectic substances are polymers of D-galacturonic acid with varying proportions of the carboxyl groups esterified with methyl groups. These polysaccharides are found in the middle lamella and in the cell walls of plant tissue. Pectin used for making jelly is obtained from slightly immature or just-ripe fruits. Most commercial pectin is derived from lemon and lime peels. A small amount is extracted from apple pomace. Sugar and acid are required for pectin to form a gel. Sufficient pectin must also be present for gel formation. The amount of pectin present in fruit juice can be determined by measurement of the viscosity of the juice or by precipitation with alcohol.

■ EXPERIMENT 1: HISTOCHEMICAL LOCALIZATION OF PECTIC SUBSTANCES

Introduction and Objective

Staining plant tissue with ruthenium red results in a red color if pectic substances are present in fairly high concentration. The basis of the reaction is unknown. The objective of this exercise is to introduce the use of microtechnique as a research tool and to localize pectic substances in a plant tissue.

Materials

- Apple — 1
- Ruthenium red (should be prepared fresh at least every semester) — 10 ml
- Straight edge razor or razor blade — 1
- Microscope and slides

Completion time: 30 minutes
Complications: Be careful when using razor

Procedure

1. Using a good straight edge razor made of steel that is not excessively brittle or a single-edged razor blade, practice cutting thin (15 to 20) sections of fresh tissue such as an apple.

2. Float one of the thinner cut sections on a drop of water on a microscope slide. Flood with aqueous ruthenium red (1:5000). Observe under the microscope for a pink to red color indicating pectic substances.

Questions

1. Where were the pectic substances observed? What role do they play in the intact plant?

2. What changes do pectic substances undergo as fruits ripen? At what stage are they useful to the food industry?

■ EXPERIMENT 2: PECTIN GELS

Introduction and Objective

Pectin, when present with sufficient sugar and at the correct pH, will form a firm gel. Pectin concentration, sugar concentration, and acid concentration all affect the nature and strength of the gel formed. It is the objective of this exercise to determine the pectin content of a juice and to determine the effect of sucrose, pH, endpoint cooking temperature, and type and concentration of pectin on percent soluble solids, yield, percent sag, and strength of pectin gels.

Formulas (See Table 18.1)

TABLE 18.1

Variations Used in the Pectin Experiments

Variations		Fresh Juice (ml)	Sugar (g)	Endpoint Temp. (°C)
A.	Vary amount of sugar			
1.	$1/2$ optimum	237		104.5
2.	Optimum	237		104.5
3.	$1^1/2$ optimum	237		104.5
B.	Vary pH of juice			
4.	Juice pH	237		104.5
5.	Adjust to pH 2.4	237		104.5
C.	Vary endpoint temperature			
6.	Low endpoint	237		101.5
7.	Optimum	237		104.5
8.	High endpoint	237		107.5

Variations		Pectin (g)	Calcium Chloride (g)	Distilled Water (ml)	Sugar (g)	Citric Acid[a] (ml)	Endpoint Temp. (°C)
D.	Vary amount of 150-g grade pectin						
9.	$1/2$ optimum	1.5	0	300	454	4.7	104.5
10.	Optimum	3.0	0	300	454	4.7	104.5
11.	Twice optimum	6.0	0	300	454	4.7	104.5
E.	Vary type of pectin						
12.	Low methoxyl	5.0	0.365	300	10	4.7	Bring to boil
13.	Fast setting (plus 50 g pureed strawberries)	3.0	0	300	454	4.7	104.5
14.	Slow setting (plus 50 g pureed strawberries)	3.0	0	300	454	4.5	104.5

[a] 286.02 g hydrous citric acid dissolved in 300 ml H_2O.

Completion time: 2 hours

Complications: Calibrate your thermometer to make sure you get the proper endpoint cooking temperature. Also, be sure that the thermometer is in the juice and not just in the froth on top of the juice. Remember to record the empty weight of the jelly glasses so you can calculate yield. Because of the number of variations, this exercise is best done by several groups of students in the lab.

Procedure

1. Preparation of juice: Wash apples, quarter, and discard stem and blossom ends. Do not pare or core. Cut into small pieces. For each 4 pounds of apples, put into a kettle and add 4 cups of water. Cover and bring to a boil on high heat. Reduce heat and simmer for 20 to 25 minutes or until tender. Stir to prevent scorching. Pour the fruit into a damp jelly bag or three layers of cheesecloth spread over a colander and allow to drip into a bowl. When the dripping has almost ceased, press the bag to obtain all the juice. Strain through a fresh jelly bag.

2. Preparation of jelly glasses: Immerse in water and boil 15 min. Record weight of each of two jelly glasses and one small beaker and label. Turn upside down on paper towel on tray.

3. Determine the temperature of boiling deionized water and adjust endpoint temperature for jelly if the boiling point varies from 100°C.

4. Perform alcohol test by adding 15 ml of fruit juice to 15 ml of alcohol in a graduated cylinder. Invert the graduate slowly and turn back. A precipitate in a mass indicates enough pectin for jelly. Pour out into a dish and lift the pectin.

5. Determine the amount of sugar to be added for jelly variations 1 through 8 (Table 18.1) by performing the jelmeter test on the juice. Refer to the Equipment Guide section for use of the jelmeter.

6. Preparation of jelly: For variations 1 through 8 (Table 18.1), cook sugar and apple juice in an aluminum saucepan over direct heat. Boil rapidly but do not boil over. Cook to the endpoint temperature given in the table. This is an indication of the sugar concentration. The thermometer must be in the juice — not in the froth. For variation 12, mix pectin with $CaCl_2$. Add pectin mixture slowly to the cold water while blending in a blender. For variations 9 to 11, 13, and 14, mix pectin with $^1/_{10}$ of the sugar. Heat the water and add the pectin mixture while stirring. For variations 13 and 14, add strawberries. For variations 9 through 14, bring mixtures to a boil for $^1/_2$ min. Add remainder of sugar and heat to the endpoint temperature given in the table. Add acid.

7. When the endpoint temperature is reached, perform the sheet test: Dip a cold metal spoon into the hot jelly. Holding it above the pan, turn it so the liquid runs off the side. The jelly is done if a couple of the drops that cling to the edge just run together and "sheet" from the spoon rather than fall as individual drops. If sheet test does not work, do not cook beyond indicated endpoint temperature.

8. Remove jelly from heat, pour immediately into jars and a small amount into a small beaker or custard cup. *Note:* If gelation is carried too far before the jelly is poured, the mechanical disruption caused by pouring decreases the gel strength. When jelly in jars is cool, rest lid on jelly glass and label with name, date, lab section, and variations.

9. Determine yield by weight. Weigh the jelly glasses and beaker plus jelly, and subtract the weight of the glasses and beaker or custard cup.

10. With a refractometer, measure percent soluble solids of jelly in beaker as soon as possible before a gel forms; results should correlate with sugar concentration. Refer to Equipment Guide section for use of refractometer.

11. To determine percentage of sugar in finished jelly:

$$\frac{\text{weight of sugar used}}{\text{weight of jelly}} \times 100 = \% \text{ sugar}$$

12. To determine gel strength of jellies, store jelly in jars until the next laboratory period and test with a cone probe and penetrometer or a Texture Analyzer. Refer to the Equipment Guide section for use of the penetrometer or Texture Analyzer.

13. To determine percentage sag, use Vernier calipers and record height of jelly in the beaker or custard cup (Reading 1) and after unmolding jelly onto a dish (Reading 2).

$$\% \text{ sag} = \frac{\text{Reading 1} - \text{Reading 2} \times 100}{\text{Reading 1}}$$

Refer to Equipment Guide section for use of Vernier calipers.

14. Rank jellies for clarity, firmness, and flavor.

15. Graph each variable against percent soluble solids, against yield, against sag, and against gel strength.

Questions

1. What physical property of a juice does the jelmeter measure?

2. Why does alcohol precipitate pectin?

3. What stabilizes pectin in a colloidal dispersion?

4. What factors in jelly-making destabilize the colloidal dispersion?

5. What function does cooking jelly to 104.5°C serve?

6. If jelly is cooked to 103°C instead of 104.5°C, how will this affect:
   - Yield of jelly?
   - Concentration of pectin?
   - Concentration of sugar?
   - Percentage sag of jelly?

7. If jelly was made with ¹/₂ optimum sugar but cooked to the same endpoint temperature as the optimum, how would this affect:
   - Yield of jelly?
   - Concentration of pectin?
   - Concentration of sugar?
   - Percentage sag of jelly?

8. If jelly was made with $\frac{1}{2}$ optimum sugar but cooked the same length of time as the optimum, how would this affect:

   - Yield of jelly?

   - Concentration of pectin?

   - Concentration of sugar?

   - Percentage sag of jelly?

9. If jelly was made from pectin that was 30% vs. 70% methylated, what would be the expected results? How might the formulas differ? Which of these would be fast setting and which slow setting?

10. If the amount of pectin in a jelly was greater than optimum, but the endpoint temperature and sugar remained the same, how would this affect:

   - Yield of jelly?

   - Concentration of pectin?

   - Concentration of sugar?

   - Percentage sag of jelly?

19 LABORATORY: SYNTHESIZED CARBOHYDRATE FOOD GUMS

Natural gums, such as guar, locust bean gum, gum arabic, or carrageenan, often lack one or more desirable properties, such as acid or thermal stability, specific viscosity, ingredient compatibility, availability, organoleptic attributes, or low cost. To achieve these properties and many others, food gums have been manufactured from cellulose and chemically modified to provide specific characteristics in end use. Some common examples are methylcellulose, hydroxypropyl methylcellulose, and sodium carboxymethylcellulose. These gums can coat, stabilize, suspend, bind, form films, carry flavors, thicken, reduce syneresis, and improve texture and shelf life. Certain natural gums such as alginate can interact with metal ions such as calcium to produce "instant" gels. Such chemistry finds uses in foods such as frozen onion rings made from diced onions and the pimento in pimento-stuffed green olives.

■ EXPERIMENT 1: DISPERSIBILITY AND THERMOGELATION OF CELLULOSE GUMS*

Introduction and Objective

Pure cellulose can be chemically modified by introducing chemical substituents such as methyl and/or hydroxypropylmethyl ether groups. These cellulose gums have a number of interesting characteristics, including the ability to stabilize emulsions and suspensions, acid stability, and the ability to undergo thermogelation (gelation on heating). The objective of this exercise is to investigate some properties of cellulose gums and to learn to make gum dispersions.

Materials

- Methocel A4M — 5 g
- Methocel K4M — 5 g
- Vinegar — 40 ml
- Anti-foam agent — 2 drops
- Vegetable oil — 50 ml
- Pepper — 0.5 g
- Flour — 60 g
- Sugar — 38 g

---

* Methocel experiments courtesy of The Dow Chemical Company, Designed Products Department, Midland, MI 48640.

Completion time: 1.5 hours

Complications: none

Procedure — Methocel Dispersion Preparation

1. Cold water trial: Add 1 g Methocel A4M Premium to 99 g water at room temperature. Stir vigorously. What happened?

2. Hot/cold technique: Make 100 g dispersions of 2% Methocel A4M Premium and 2% Methocel K4M Premium by dispersing Methocel powder (2 g) with vigorous stirring in one-third to one-half of the required total volume of deionized water as hot water (85–90°C). Add the remainder of the water (to bring to 100 g total) as cold water or ice. Continue to agitate until the mixture is smooth and the temperature has reached 22°C. *Label and save these solutions for later use.*

3. Nonaqueous dispersion

   a. Oil: Disperse 1 g Methocel K4M Premium into 38 g salad oil with mild agitation; add 23 g water to hydrate the gum particles; add 38 g vinegar and stir. For comparison make a similar mixture of oil, water, and vinegar without carbohydrate food gum. Note the relative times for separation. *Save these preparations for later use.*

   b. Dry-blending: Add 0.4 g Methocel K4M Premium to 100 g batter mix (60 g flour, 38 g sugar, 2 g salt) and stir-blend 2 min to separate the Methocel particles from each other. Add 120 g cold water and mix gently until the batter is smooth. Does this method allow adequate hydration of the carbohydrate food gum?

Procedure — Methocel Properties

1. Thermal gelation: Half-fill an 18 × 150 mm test tube with 2% solution of Methocel A4M Premium (*from the hot/cold technique step above*). Prepare a second tube with 2% Methocel K4M Premium (*from the hot/cold technique step above*). Place both tubes in boiling water. Watch the rate at which each gel forms. Which forms first? After thermal gelation has occurred, carefully slide the gel out into a watch glass or Petri dish. Feel the texture of the gels, note the color, and observe the relative rates of relaxation. Which gel is not firm? Which gel disappears soonest?

2. Emulsifier: Reexamine the emulsions prepared in Step 3 (nonaqueous dispersion in oil) above. In addition to the relative times of separation, note the color, texture, and taste of each.

3. Low pH stability: Determine the pH values of the oil, water, and vinegar mixtures (*from Step 3*) both with and without the food gum emulsifier.

4. Suspension aid: Add a small amount (tip of spatula) of black pepper to each of the oil, water, vinegar mixtures (*from Step 3*) and shake. Observe after 5 minutes. Which mixture appears to have the most uniform dispersion of the black pepper?

5. Surface activity: Dilute 12.5 ml of 2% Methocel A4M Premium (*from the hot/cold technique step above*) to 100 ml with water and shake vigorously. Note the stability of the foam. Add one drop of food-approved anti-foam agent. Note action. Shake again. Did the foam recur?

6. Odor/taste: Dilute 25 ml of 2% Methocel K4M Premium (*from the hot/cold technique step above*) to 100 ml with deionized water and mix thoroughly. How do you judge the sensory level of this 0.5% gum solution?

   a. Odor: (circle one) strong — moderate — weak — none

   b. Taste: (circle one) strong — moderate — weak — none

Evaluation and Discussion Questions

1. Suggest reasons why the thermal gelation property is useful in various food products (deep-fried foods, cake yield improvement, pie filling yield)?

2. Explain the reasons why the various dispersion techniques work better in comparison to the cold water dispersion. What are "fish eyes"? Name other thickening agents where these techniques would be helpful.

3. Name three products and the principles demonstrated over the semester where carbohydrate gums might be appropriate for use.

■ EXPERIMENT 2: ALGINATE GELS

Introduction and Objective

Gums in foods come from a variety of sources. They can be material exuded from plants, endosperm material of plant seeds, cellulose derivatives, or bacterial in origin. Some of the more interesting gums come from seaweeds. One of these is alginate, which is derived from the giant kelp, *Macrocystis pyrifera*. Alginates have very interesting interactions with divalent cations, especially calcium. The objective of this exercise is to study gel formation and the characteristics of alginate gels.

Materials

- Low calcium sodium alginate (Keltone algin) — 1 g
- $CaCl_2$ — 5 g
- Food coloring

Procedure

1. Prepare 100 ml of a 5% $CaCl_2$ solution.

2. Prepare 100 ml of a 1% alginate dispersion by adding water to the gum through the hole in the top of the blender cup lid while blending. Add food coloring.

3. Pour the gum dispersion slowly into the $CaCl_2$ solution and observe instant gel formation.

4. Remove strands of gum and place on a watch glass. Observe syneresis over several days.

Questions

1. What is the mechanism of gel formation?

2. How does this gel compare to low methoxyl pectin gels? To hydrogen-bonded gels?

20 EQUIPMENT GUIDE

■ BROOKFIELD VISCOMETER (ANALOG AND DIGITAL)

Measures

Viscosity by measuring the force required to rotate a spindle in a fluid.

Uses

Viscosity of fluids for quality control, or to study rheological properties of a test material since the rate of shear can be varied.

Procedure (Analog)

1. Attach spindle (Figure 20.1A) to lower shaft. Lift the shaft slightly while holding it firmly with one hand. Screw in the spindle by turning to the left (clockwise).

2. Depress clutch (C) and insert spindle in the test material until the level of the fluid is at the immersion groove cut in the spindle shaft. With a disc-type spindle, tilt the spindle when immersing it to avoid trapping air bubbles. Avoid hitting the spindle against the side of the container while it is attached to the viscometer since this can damage the shaft alignment.

3. Level the viscometer using the bubble level (B) as an aid.

4. Depress the clutch (C) and turn on the motor switch (D). Turn the speed control knob (E) to the desired speed with the motor running. Release the clutch and allow the dial to rotate until the pointer (F) stabilizes at a fixed position to the dial (20 to 30 sec). Depress the clutch (C) and snap the motor switch (D) to stop the instrument with the pointer in view.

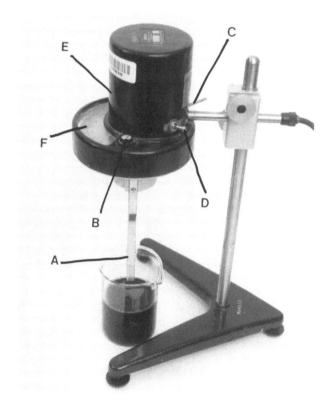

Figure 20.1 Analog Brookfield viscometer.

5. Find the viscosity in centipoise (cP) with the Factor Finder using the appropriate viscometer model type and spindle number and speed.

6. To obtain replicate readings, hold the original reading with the clutch still depressed and start the viscometer. Release the clutch. After the pointer stabilizes, depress the clutch and take another reading. This will reduce the time required to stabilize the pointer. Readings should be reproducible to within 0.2% of the full scale under constant temperature, and accuracy is guaranteed to within 1% of the full scale range.

7. Record data in the following format:

Sample	Model	Spindle no.	r/min	Dial Reading	Factor	Viscosity (cP)	Temperature (°C)

Procedure (Digital: RVDV-E Brookfield Viscometer)

1. Turn on power to viscometer. (Use rocker switch on back; see Figure 20.2.)

2. Select a spindle and attach it to the lower shaft. Set the SPEED/SPINDLE switch in the right position and rotate the SELECT knob until the spindle number displayed on the screen matches the one you attached to the viscometer.

3. Insert and center the spindle in the test material until the fluid's level is at the immersion groove on the spindle's shaft.

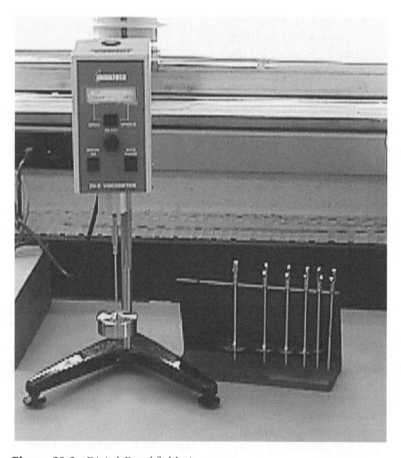

Figure 20.2 Digital Brookfield viscometer.

4. To make a viscosity measurement, select a speed. To do this set the SPEED/SPINDLE switch in the left position and rotate the SELECT knob until the desired speed is selected.

5. Switch the motor switch to the ON position. Allow time for the indicated reading on the screen to stabilize.

■ COMPENSATING POLAR PLANIMETER

Measures

Plane areas of any shape.

Uses

The area of a slice of baked product is used as an index to the volume of the product. This method of assessing volume has the advantage over seed displacement that samples do not have to cool before the measurement can be made.

Procedure

1. Place a piece of paper that will lie perfectly flat on a flat surface and attach it with tape or thumbtacks.

2. Set up the tracer arm (Figure 20.3A) a little to your right and perpendicular to the front of the table. Insert the pole arm ball in its socket (B) on the tracer arm.

3. Holding the magnifying glass (C), test the limits of the motion of the tracer point from side to side and note that it can be moved freely whenever the tracer arm and pole arm make an angle within the wide arc between about 15 and 165°.

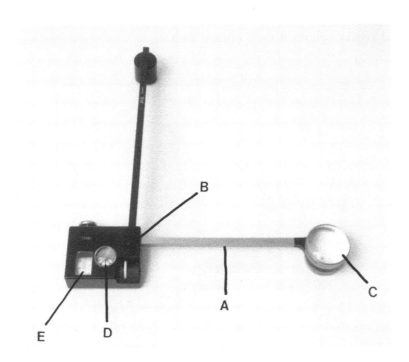

Figure 20.3 Compensating polar planimeter.

4. Put the surface of the sample on the paper and trace around it. Draw a fine line across the edge of the figure to mark the starting and finishing point. Put the tracer point in the magnifying glass exactly at this line and read the planimeter or set it to read zero. Holding the magnifying glass (C), follow the outline of the figure carefully in a clockwise direction. Avoid any counterclockwise movement. If the tracer point leaves the line slightly, the error can be compensated for by moving the tracer an equivalent amount away from the line in the other direction. When the tracer has returned to the starting point, make a second reading. The difference between the two readings is the area of the figure. To make readings: The first digit of the reading is the number on the dial (D) at which the pointer is standing or which it has just passed. The second and third digits are read from the measuring wheel (E) by observing the line that is opposite or has just been passed by the zero on the Vernier scale. The second digit is the number or major division on the measuring wheel, and the third is found from the single divisions on the measuring wheel. The last digit is the graduation on the Vernier that makes a straight line with a wheel graduation. One Vernier unit equals 0.1 square centimeters (cm^2). A simple division on the wheel equals 1.0 cm^2 and a major division equals 10 cm^2. One division on the dial equals 100 cm^2.

■ CONSISTOMETER (BOSTWICK)

Measures

Viscosity as reflected by the flow of a product under its own weight during a specified time.

Uses

Consistency of viscous materials such as catsup, jellies, preserves, baby foods, and salad dressings.

Procedure

1. Place the consistometer on a level surface and adjust the leveling screws (A) until the bubble in the spirit level (B) is centered (see Figure 20.4).

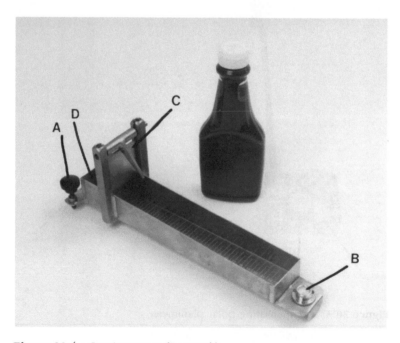

Figure 20.4 Consistometer (Bostwick).

2. Close the gate (C) to the reservoir and hold it down by hooking the trigger in place. Fill the reservoir (D) with the product to be tested and level off the top.

3. Press down the trigger to release the gate and start timing. At the end of the selected time period, average the readings at the center of the trough and at the edge of the trough. Time of flow and temperature of the sample should be controlled for comparison of readings.

■■ HUNTER COLORIMETER

Measures

Color. This instrument can be used to measure the color of a variety of products. Its unique geometry "sees" color the way the human eye sees color, and it can report color in a variety of ways, including the common L, a, b Hunter parameters.

Uses

Measuring color of baked goods, fruits, vegetables, meats, and a variety of other types of food.

Procedures

1. Turn on the Lab Scan XE (right rear of the unit at bottom under cord plug-in; see Figure 20.5).

2. Turn on the computer and monitor.

3. Double click the UNIVERSAL icon on the desktop.

4. Standardize the instrument:

   • Choose Sensor, then Standardize on the tool bar.

   • When the Standardize dialog box appears, click OK.

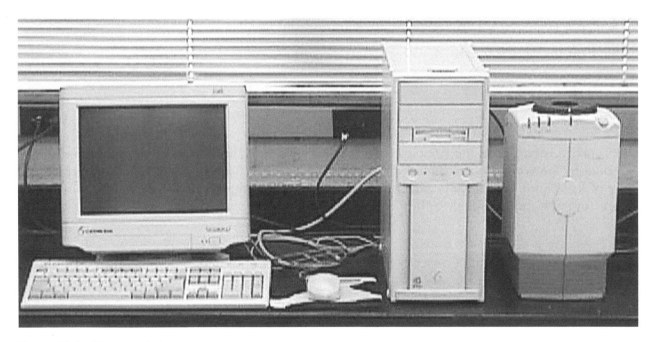

Figure 20.5 Hunter colorimeter.

- Place the Black Glass at the measuring port; click OK.

- Place the White Tile at the measuring port; click OK.

- You should get a message saying "Sensor Successfully Standardized"; click OK.

5. Place your sample at the measuring port in a Petri dish. Be careful to not let food fall into the Lab Scan XE.

6. Click the Read Sam (Sample) button on the toolbar.

7. The Average Hunter Lab box should appear.

8. Click "Average."

9. Name your sample in the Enter Sample Information box. *Note*: Each sample you measure must have a unique ID.

10. Click OK.

11. You should now see your data in the Master Color Data window.

12. You can easily change the active view to a different color system through the "Active View" button on the toolbar.

13. Click the active view button and change "Scale" to (say) Yxy, click OK.

14. The Master Color Data window now shows the color data in Y, x, and y values as shown.

15. Record your data from the Master Color Data window.

16. Close the Universal main window.

17. You will see a box asking if you want to exit without saving unsaved readings.

18. Click "Exit."

19. Shut the computer down via the Start button on the taskbar.

20. Shut off the Lab Scan XE.

■ HYDROMETER

Measures

Specific gravity is based on the principle that a solid suspended in a liquid will be buoyed up by a force equal to the weight of the liquid displaced. The weight of the displaced liquid is equal to the product of its volume and density, and the volumes of different liquids displaced by the same floating body are inversely proportional to the densities of the liquids.

Uses

Sugar, salt, or alcohol content of aqueous solution; solids content of milk or tomato juice; identification of oils.

Procedure

1. The weight of the hydrometer must be less than that of the liquid it displaces. Place liquid to be tested in a glass cylinder at least twice as wide as the diameter of the bottom of the hydrometer (see Figure 20.6). The liquid should be free of air bubbles and at the temperature specified on the hydrometer.

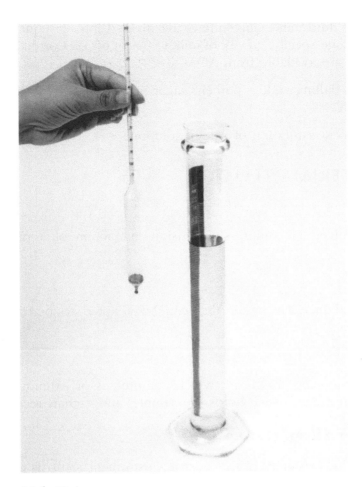

Figure 20.6   Hydrometer.

2. Slowly immerse clean, dry hydrometer in the liquid, slightly below the point where it floats, and then allow it to float freely.

3. Take reading with the line of vision horizontal to the meniscus. The point is taken where the surface line cuts the hydrometer scale.

4. The stem is calibrated by different scales according to the type of hydrometer.

Types of Hydrometers

- Alcoholometer — percent alcohol by volume.

- Baume hydrometers — read in degrees Baume (Be°). Degrees Baume can be converted to specific gravity (sp gr) as follows.

  For liquids heavier than water:

  $$\text{s.g.} = \frac{145}{145 - \text{Be}°}$$

  For liquids lighter than water:

  $$\text{s.g.} = \frac{140}{130 + \text{Be}°}$$

- Lactometers — for fluid milk. Quivenne: scale divided into 25 equal parts from 15 to 40; 29 represents the average specific gravity of milk (1.029 at 60°F). Correction tables for conversions to specific gravity are available from AOAC.

- Saccharometers — Balling scale = percent sugar by weight at 60°F; Brix scale = percent sugar by weight at 17.5°C.

- Salometers — percent saturation of salt in solution.

■ INSTRON MATERIALS TESTER

Measures

Force and/or energy involved in compression/tension interaction of a probe with food or other materials.

Uses

The Instron is a universal materials tester that can be adapted to perform various food texture evaluations.

Description

The Instron consists of a moving crosshead, which compresses or extends the sample; a load cell, which measures force; a test cell, which holds the sample; and a chart recorder or computer.

Procedure for Kramer Shear Testing

1. Setup: Because of the versatility of the Instron, instrument setup is a complicated procedure. You can consult the operator's guide for information on load cell calibration and selection, crosshead and chart speed adjustments, etc. The Instron requires approximately $1/2$ hour warm-up time.

2. Testing:

   a. If the sample is fresh or whole, cut to appropriate size or into chunks. If the sample is canned, drain thoroughly.

   b. Weigh 100 g of sample into the test cell. Replace the cap so that sides with stamped numbers are facing the same direction.

   c. With crosshead in the up position, slide sample compartment into test cell holder with the beveled edges of the cap pointed towards the blades. *Important:* Be sure that the sample compartment is positioned correctly (blades snug against pegs in cap) or damage to the load cell may result. Tighten sample cell holder knobs.

   d. Place selector knob in the "Return" position. Check that the chart power switch is in the "Test" position and that the pen is turned on.

   e. Depress the "Down" switch. The chart will start automatically. The test is completed when the crosshead returns to the "Up" position.

   f. Peak force (i.e., toughness) can be measured directly from the graph or can be displayed on the data integrator by keying the sequence "DISPLAY PEAK FORCE." Total energy (force through the distance of crosshead travel) can be calculated by measuring the area under the peak or by keying the sequence "DISPLAY ENERGY."

▬ JELMETER

Measures

Rate of flow.

Uses

Indication of the amount of pectin in fruit juice, which is used to determine the amount of sugar needed to make jelly.

Procedure

1. Hold finger under the bottom of the Jelmeter (see Figure 20.7).

2. Pour in the juice to be measured. It must be at room temperature. Fill to the top.

3. Remove finger from the bottom of the Jelmeter and allow the juice to flow for exactly 1 min. Replace finger.

4. Note the level of juice and the figure nearest this level, which indicates the cups of sugar to add for each cup of juice.

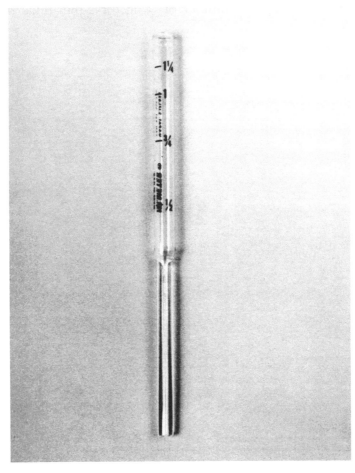

Figure 20.7   Jelmeter.

▬ LINESPREAD APPARATUS

Measures

Consistency of foods in terms of their ability to spread on a flat surface.

Uses

Apparent viscosity of foods that have fluidity such as sauces, soft custards, applesauce, starch puddings, cake batter, cream fillings, and creamed-style corn.

Procedure

1. Center glass plate above grid with numbered concentric circles at $1/8$-in. or 2-mm intervals (see Figure 20.8). Surface must be level.

2. Place the hollow metal cylinder having a diameter of 2 in. directly over the smallest circle.

3. Fill the cylinder with the food to be tested and level off with a spatula.

4. Lift the cylinder and allow the food to spread. For accurate comparisons of results, the temperature of the sample and the time of flow should be constant.

5. Quickly take readings at four widely separated points of the limit of spread of the food. Report the average of the four readings in number of $1/8$-in. or 2-mm units from the cylinder at that temperature and time of flow.

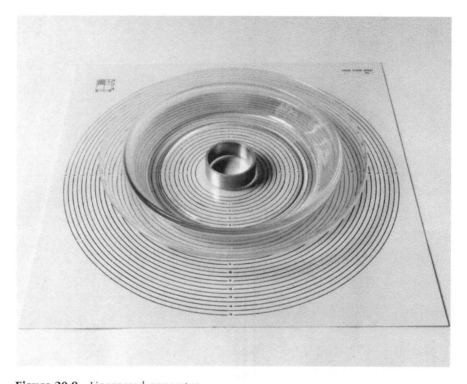

Figure 20.8 Linespread apparatus.

PENETROMETER OR COMPRESSOMETER

Measures

With the needle-like probe attachment, penetration is measured. With the cone attachment, compression and shear are measured. With the flat disc attachment, compressibility is measured if the sample is smaller in circumference than the disc. If the sample is larger than the disc, shear force and compressibility are measured.

Uses

Probe attachment — firmness or tenderness of fleshy plants or meat. Cone attachment — rigidity, plasticity, or firmness of fats, cheese, gels; degree of softness of cakes and breads; binding powers of dried milk. Flat disc attachment — relative hardness of baked products and changes in firmness in staling of bread.

Procedure

1. Level the penetrometer using the adjusting screws on the base (Figure 20.9A). Be sure to look from directly overhead when adjusting the spirit level. Weights may be added to test rod (B) if extra weight is needed to penetrate a stiff sample. The weight of the standard test rod is 47.5 g.

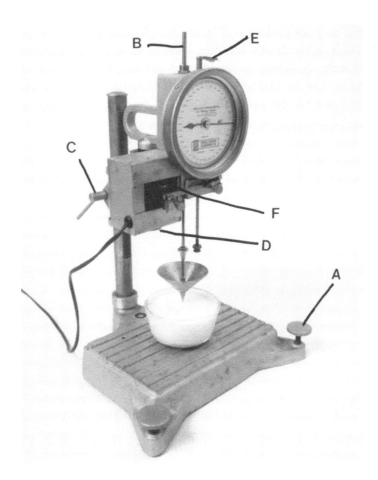

Figure 20.9 Penetrometer or compressometer.

2. Set dial reading to zero by grasping the test rod (B), depress the thumb release lever (F), and raise the test rod as high as it will go. If the dial reading is not exactly zero, adjust the zero adjusting nut.

3. Adjust the height of the mechanism head to bring the point of the penetrating instrument exactly into contact with the surface of the sample. This adjustment may be accomplished more easily by placing a weak light to one side of the sample container and following the shadow formed by the penetrating cone tip on the surface of the sample until no light appears from its point. To adjust the height of the mechanism head, first release the lock screw (C). Make a coarse adjustment by means of the coarse adjusting screw (knob directly opposite lock screw on mechanism head). Be sure to lock the head securely by means of the lock screw. Make the final "contact" adjustment by means of the micrometer adjusting screw (D). Turn on the line switch of the timer control box. (Leave its control switch in the off position at this time.)

4. To begin the test, turn on the control switch any time the pilot is out. When the pilot light goes on, the solenoid will be activated and the penetration begun. The light will stay on for 5 sec and then go out. At this time, the solenoid will become deactivated and end the penetration and the test. Turn off the control switch and take the penetration reading.

5. Reading penetration: To read the depth of penetration, push down the depth gauge (E) rod gently — as far as it will go. The dial reading now indicates the depth of penetration directly in tenths of millimeters. For example, if the pointer comes to rest at the fourth mark past the 270 point, the depth of penetration is 274 tenth-millimeters or 27.4 millimeters. The Precision Universal Penetrometer with Timer enables penetration measurements to be made to a total depth of 62 mm on a single reading. On depths greater than 38 mm, the dial pointer makes a complete revolution and moves past the zero position for a fraction of another revolution. Add the fractional revolution to determine the total depth of penetration.

6. Reset the dial reading to zero as in Step 2.

7. The coarse adjustment can be left for any given series after the original adjustment, and the fine adjustment screw may be used for the final "contact" of individual samples. Test at least three samples or three different readings for the same sample.

■ pH METER

Measures

pH.

Uses

To determine the hydrogen ion concentration of a sample.

Procedure

1. pH meters should be left plugged in and turned on all the time (see Figure 20.10).

2. Operation knob (button) should be on standby and electrodes stored in a 250-ml beaker with deionized water.

3. pH meter should be "calibrated" before use and recalibrated after five or six readings. Fresh standard solutions should be used for calibration at the beginning of each lab and thrown out when the lab is over. The solution need not be changed during the lab period unless it becomes contaminated.

Figure 20.10   pH meter.

4. To zero:

   a. Turn the temperature compensator to the temperature of the buffer being measured. Check to see if the test solution temperature is different.

   b. Carefully rinse electrodes with deionized water.

   c. Wipe off with Kimwipes.

   d. Resubmerge electrodes in a 50-ml beaker that is half full of a standard buffer with a pH of 7. Do not let the electrodes touch the bottom or sides of the beaker. Submerge electrodes carefully.

   e. Turn operation knob (or push button) from "standby" to "pH." Digital display should read a pH of 7 but will probably need to be standardized. In order to standardize, turn "calibrate" knob until the display reads exactly 7.

   f. To make sure that the pH meter is working correctly, the electrodes must be resubmerged in another 50-ml beaker with a pH standard of either 4 or 10.

g. Return pH meter to "standby." Lift the electrodes out of the pH 7 buffer, move the 50-ml beaker, and rinse electrodes with deionized water from a squeeze bottle. Dry with Kimwipes and resubmerge in the pH standard of 10 (or 4). Move operation knob from standby to pH. Recalibrate, if necessary, to the appropriate pH. If the display registers far away from 10 (or 4), see the instructor as the pH meter may be malfunctioning.

5. To measure pH of test solution after zeroing:

   a. Turn the meter to standby.

   b. Remove electrodes, clean, and dry as before.

   c. Resubmerge in an adequate amount of test solution, turn pH meter to "pH," wait 30 sec and record the reading.

   d. If the display "drifts" over a wide range, check with the instructor, as the pH meter may not be working correctly.

6. To store electrodes between uses, thoroughly clean and dry as before. Resubmerge in deionized water.

■ REFLECTANCE METER (PHOTOVOLT)

Measures

Reflectance from a sample surface.

Uses

Color or lightness comparison of food products such as coffee, potato chips, peanut brittle, cocoa, catsup, and others.

Procedure

1. Turn on the power switch (see Figure 20.11).

2. Plug the search unit into the socket on the rear apron. Position search unit as it will be for use with samples.

3. Set the meter to zero with the lamp switch off by turning the amplifier zero control until the scale reads zero. Turn on the lamp switch.

4. Depress the lamp switch to turn on the search unit, and allow it to warm up for about 30 min.

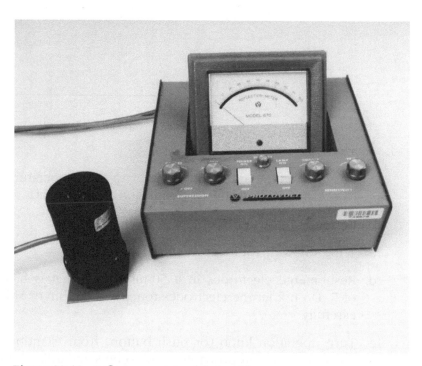

Figure 20.11   Reflectance meter (photovolt).

5. Insert the green tristimulus filter, and place the calibrated standard plaque against the search unit opening. Then set the meter pointer to the value shown on the standard plaque for the green filter by adjusting sensitivity controls on the right side of the instrument.

6. Place the search unit opening against the sample and read the meter. This represents the green stimulus reflectance reading of the sample. If lightness rather than hue is of interest, this reading is sufficient.

7. If hue is of interest, repeat the above procedure using the blue and amber tristimulus filters, and obtain the blue and amber reflectance readings of the sample. If desired, the blue, amber, and green reflectance values can be converted to C.I.E. tristimulus values according to the following formula:

$$\frac{0.8 \text{ amber reading } + \text{ } 0.18 \text{ blue reading}}{100} = X$$

$$\frac{\text{green reading}}{100} = Y$$

$$\frac{1.18 \text{ blue reading}}{100} = Z$$

These values can be further converted to chromaticity coordinates as follows:

$$x = \frac{X}{(X + Y + Z)}$$

$$y = \frac{Y}{(X + Y + Z)}$$

$$z = 1 - (x + y)$$

Use x and y coordinates to plot on chromaticity diagram on p. 7.

■ REFRACTOMETER (ABBE)

Measures

The angle to which light is bent (refracted) by a substance.

Uses

Soluble components such as the soluble solids content of sugar syrups and fruit products and the degree of hydrogenation of fats.

Care of the Refractometer

1. The refractometer must be kept scrupulously clean at all times. Dust, oil, and solid materials, if allowed to accumulate on any part of the instrument, will find their way into the bearings and hinges causing wear and eventual misalignment. The operator should make it a practice at the close of each day's work to clean all exposed surfaces thoroughly.

2. The prism should be thoroughly cleaned after each test and should be kept closed when not in use. In this type of instrument, the glass of which the prisms are made is of high refractive index and inherently soft. It is therefore easily damaged by surface scratching and corrosion. If a dust film is allowed to accumulate on the polished surface, its removal can cause more

damage than many hours of actual service. The gradual deterioration of surface quality results in hazy borderlines; hence, every care should be exercised to protect and preserve the prism surfaces.

3. Prisms should always be cleaned immediately after use. Where possible, wipe first with a clean, dry lens tissue followed by a tissue or cotton swab dampened with water, alcohol, or other suitable solvent (not acetone). Never use a sharp object such as a knife, needle, etc., on either the prism or the seal around the prism. Even a slight crack in the sealer may cause serious damage to the prism mounting, which will necessitate considerable repair. Do not dry the surfaces by rubbing with cotton. Lens tissue, if kept in a closed container, may be employed if used lightly. Thoroughly washed linen may also be safely used. Avoid the use of any cleaning material, either linen or tissue, that has been lying on the work table where it can pick up dust or grit.

Procedure

1. Turn on instrument at switch on cord.

2. A liquid at a constant temperature may be circulated through prism housings if necessary.

3. Introduce a drop of sample between the upper and lower refractive prisms (Figure 20.12A). Adjust the light position so the maximum intensity can be seen through the ocular.

4. After the sample is in position on the instrument, set the scale at the approximate value expected. To see the scale, depress the momentary contact switch (B).

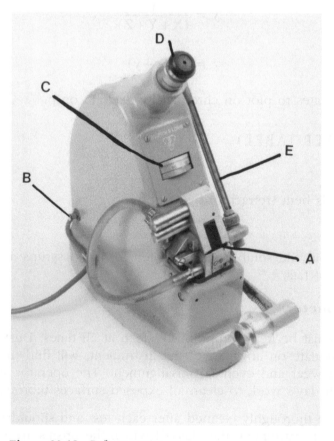

Figure 20.12   Refractometer.

5. Release the switch and bring the borderline, which will probably be strongly colored, near the crosshair and compensate the color by adjusting the position of the dial (C). The borderline should be faintly blue on one side and faintly red on the other.

6. Observe the crosshairs, sharply focusing the eyepiece (D) if necessary, and bring the dividing line upon their intersection by means of the coarse or fine hand controls (on right side of refractometer).

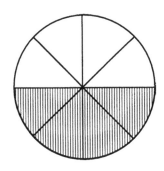

7. Read the refractive index and total scales by depressing the momentary contact switch (B). Estimate the refractive index scale (upper) to the fourth place. The total soluble solids scale (bottom) reads directly to 0.2% and can be estimated to 0.1%.

8. If working with liquids, record both index and the prism temperature at the time of reading. If your sample is not at 20°C, use the correction thermometer (E) and add or subtract from the scale reading or refer to the temperature correction table in the instruction manual.

■ SEED VOLUME APPARATUS

Measures

Volume.

Uses

The volume of baked foods such as cakes, bread, biscuits, and muffins.

Procedure

1. Open gate (A) between column and lower container (see Figure 20.13). Turn the column into an upside-down position to allow seeds to flow into upper reservoir. Close the gate and turn column upright.

2. Unlatch (B) lower container and tilt column to open the container.

3. If the volume of a loaf of bread is to be assessed, place the 900 cm³ wooden dummy loaf standard into the container. If the sample is not a loaf of bread, do not place dummy loaf in the container. Latch the container.

4. Open the gate with a quick, smooth motion and allow the seeds to fill the empty space around the dummy loaf.

5. Read the volume on the column (C). If the seed height registers at 900 cm³, the amount of seeds to be used for testing is correct. If the column does not read 900 cm³, correct it by opening the lid on the reservoir and adding or subtracting seeds.

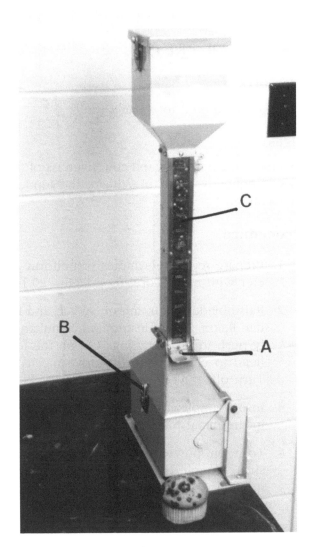

Figure 20.13 Seed volume apparatus.

6. Leaving the gate open, tilt the volume meter to the upside-down position to allow seeds to flow into the reservoir. Slap the column to remove clinging seeds. Close the gate and return the volume meter to the upright position.

7. Wrap product sparingly in clear plastic wrap so that wrap is molded to the contours of the sample.

8. To test the product, unlatch the container and remove the dummy loaf. Replace it with the wrapped sample. The test loaf should represent a sample larger than 900 cm^3. If not, seeds will have to be filled to a greater height with the dummy loaf and calculations adjusted accordingly. Relatch the container.

9. Open the gate with a smooth, quick motion and read the volume on the scale. If the sample is a loaf of bread, the volume may be recorded directly from the column in cubic centimeters. If the sample is not a loaf of bread and the seed height was adjusted to 900 cm^3 with no dummy loaf present, the volume of the sample will be the reading taken with sample minus 900 cm^3.

10. Turn the column into an upside-down position to allow the seeds to flow back into the reservoir. Slap the column to remove clinging seeds.

11. Close the gate and return the column to its upright position. The sample may be removed at this time.

▬ SHEAR PRESS

Measures

The force required to shear a food

Uses

Estimates the hardness and cohesiveness of foods; most frequently used to estimate the tenderness of meats.

Procedure

1. Prepare replicate samples of uniform diameter with metal corer (see Figure 20.14).

2. Raise blade by turning on switch and lifting handle on right of instrument. Simultaneously push in on metal lugs (A) on the testing head to engage it to the drive mechanism. Turn off power.

3. Insert sample core across triangular opening in the metal plate (B).

4. Turn on power to lower testing head to shear sample. Record pounds pressure required to shear sample by reading maximum recording needle (C).

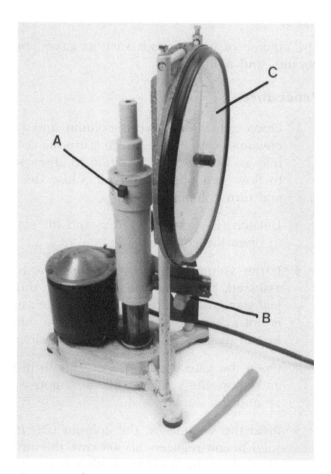

Figure 20.14 Shear press.

▄ SHORTOMETER*

Measures

Breaking strength.

Uses

Tenderness/crispness of baked products such as pastry, cookies, and crackers.

Procedure

1. Prepare samples of uniform dimensions if comparisons are to be made. Samples must be made approximately $2\frac{1}{2}$ in. long to fit across the platform.

2. Turn on instrument by pressing red button on back panel (see Figure 20.15). Adjust potentiometer on back panel until (OFFSET) (OK) is displayed.

3. Position sample across the parallel supporting bars.

4. Press "RUN" when (INPUT?) is displayed and hold until (RUNNING) is displayed.

5. Read display (FORCE =) and (BREAKING FORCE – SPECIMEN WEIGHT) in grams.

6. Press and hold "RESET" until "RESTART" is displayed and the overarm moves to the "TOP" position.

7. Turn the shortometer off when not in use — never turn the shortometer on unless it has been off for at least five seconds. *Caution:* Bars and pan must be clean. Grease and crumbs add to the breaking strength.

Figure 20.15 Shortometer.

---

* Computer Controlled Machines, Model 602.

■ SPECIFIC GRAVITY OF SOLIDS

Measures

Specific gravity by weight of known volume.

Uses

Measure amount of air incorporated in products such as whipped cream, egg white foams, creamed shortening, and cake batters.

Procedure

1. Weigh a dry container to the nearest gram.

2. Fill container with cooled, boiled deionized water at room temperature. Complete fill on balance; judge at eye level. Weigh to nearest gram.

3. Fill dry cup with test material. Do not pack. Remove excess with spatula. Wipe outside of container. Weigh to nearest gram.

4. Calculate the specific gravity as follows:

$$\text{specific gravity} = \frac{\text{weight filled container} - \text{weight container}}{\text{volume container}}$$

where volume container = (weight container + water) – weight container. Since specific gravity is the density of a substance relative to water it has no units.

■ SPECTROPHOTOMETER

Measures

The absorption of light at a particular wavelength by the sample.

Uses

Qualitative identification and quantitative determination of colored substances.

Procedure

1. Turn on instrument by rotating the left knob clockwise and allow to warm up for 15 to 30 min (see Figure 20.16). Turn wavelength dial to appropriate setting.

2. Zero instrument with cuvette chamber empty and lid closed by adjusting left knob until needle is at 0% transmittance. Insert cuvette containing reagent or tissue blank and adjust right knob until needle reads 100% transmittance. Each time the wavelength is changed, zero the instrument again.

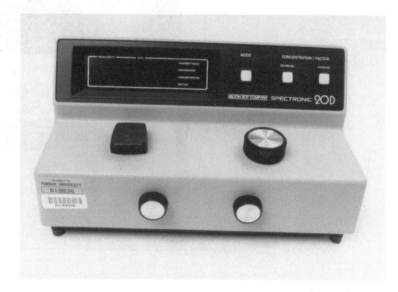

Figure 20.16 Spectrophotometer.

3. Insert cuvette with sample into chamber matching notch on cuvette and chamber.

4. Close lid and record the absorbance (optical density). *Caution*: Use clean, matched cuvettes. Remove fingerprints with Kimwipes.

■ STABLE MICRO SYSTEMS TEXTURE ANALYZER

Measures

Force involved in compression/tension interaction of a probe with any kind of food.

Uses

Texture analysis of foods.

Description

The Texture Analyzer (T.A., see Figure 20.17) consists of a moving crosshead that compresses or extends a food sample, a load cell that measures force, various probes for testing food texture, and a digital readout or computer.

Procedures

Choosing Probe Type

Select appropriate probe as follows:

- Cylinder — for flat surfaces, tackiness
- Cone — for hardness, penetration, spreadability of soft samples
- Puncture — for hardness of skins, layers
- Knife — for breaking strength, cutability

Running a Test

- Turn on the computer, monitor, and texture analyzer. (The switch is on the back left-hand side of the machine.)

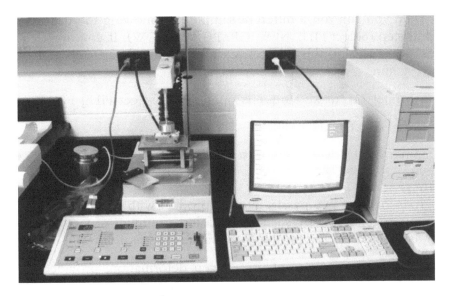

Figure 20.17 Texture analyzer.

- Select the Texture Expert software from the Windows screen.

- Select User and enter password (if any) to access program.

- Choose probe type.

- Attach the probe. Probe height can be changed by using the arrow and fast buttons on the texture analyzer. There are "safety" stops on the arm of the texture analyzer. You may get an error message if the probe goes beyond these settings.

- You can adjust the "safety" stops by manually moving them up and down.

- Prepare samples. All samples to be tested and compared should be of uniform size and shape.

- Select FILE, NEW, GRAPH WINDOW.

- Select T.A., T.A. Settings, Load.

- Choose a setting consistent with the product that you are testing.

- Choose UPDATE — This sends the T.A. settings to the texture analyzer.

- Place the sample under the probe.

- Select T.A., Quick Test Run (test will run and a graph will appear on the screen).

- Analyzing the graph: Select Process Data, Macro, Run. This option will give you various results about the graph depending on what kind of macro you have constructed. Some of the parameters you can get by doing this are peak force, area under the curve, etc. Another way to run the macro is to go to the upper right-hand corner of the screen where there is a box that lists the macros available and click the button to the left of the box to run the macro selected.

- To view only the graph you are interested in, select VIEW, GRAPH, VIEW SELECTED ONLY.

- At this point it is a good idea to write down your results. Alternatively, the Graphs and Results can be printed and saved.

- After each series of runs, close the graph (Test) window. Then close the Results window. In each case you will be asked if you want to save the results, to which you should generally answer NO. Then you can run a different sample by following the instructions above starting from the eighth step (Select FILE, NEW, GRAPH WINDOW). If you don't change sample types between runs, you can skip the Select T.A., T.A. Settings, Load step and go directly to Select T.A., Quick Test Run.

- To print, turn printer on, make sure it is online, and select FILE, PRINT.

- To exit and close, select FILE, EXIT.

- Turn off computer hard drive, monitor, texture analyzer, and printer.

Texture Profile Analysis

A texture profile analysis involves two passes into the product with a user-definable pause in between each pass. From the curve generated by such a test, a large number of factors can be determined to provide an accurate assessment of the product's characteristics as follows.

- *Hardness:* The force necessary to attain a given deformation; provided as the final peak of the texture profile analysis (TPA) curve.

- *Cohesiveness:* The quantity necessary to simulate the strength of the internal bonds making up the body of the sample. If Adhesiveness < Cohesiveness, the probe will remain clean, as the product has the ability to hold together.

- *Springiness:* The rate at which a deformed sample goes back to its undeformed condition after the deforming force is removed. This can also be called Elasticity.

- *Adhesiveness:* The quantity necessary to simulate the work to overcome the attractive forces between the surfaces of the sample and the surface of the probe with which the sample comes into contact. If Adhesiveness > Cohesiveness, then part of the sample will adhere to the probe.

- *Fracturability:* The force at which the material fractures (height of first significant break in the peak of TPA curve); a sample with a high degree of hardness and low cohesiveness will fracture. This can also be called Brittleness.

- *Chewiness:* The quantity to simulate the energy required to masticate a semisolid sample to a steady state of swallowing (Hardness/Cohesiveness/Adhesiveness).

- *Gumminess:* The quantity to simulate the energy required to disintegrate a semisolid sample to a steady state of swallowing (Hardness/Cohesiveness).

■ VERNIER CALIPER

Measures

Length.

Uses

Percent sag of gels (index to gel strength).

Procedure

1. To determine the percent sag of a gel, readings of height are taken in the container and after the gel is turned out. All comparisons should be made at the same temperature. Insert the Vernier caliper extension (A) vertically into the center of the gel (see Figure 20.18).

2. Mark the depth of insertion by pushing the Vernier caliper housing (B) down until it contacts the surface.

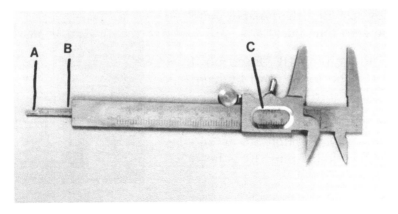

Figure 20.18   Vernier caliper.

3. Withdraw the caliper and read directly on the Vernier caliper (C).

4. To read the Vernier caliper in English units, use the top part of the scale. The rule is divided into intervals of $^1/_{16}$ in., and the sliding jaw or Vernier has eight divisions in a length corresponding to seven divisions on the rule. Thus each division on the Vernier has a length of $^7/_{16} \times ^1/_8 = ^7/_{128}$ in. The difference between one division on the rule and one division on the Vernier is $^1/_{16} - ^7/_{128} = ^1/_{128}$ in. Note the point at which a line on the Vernier lines up with a line on the rule. Add the number of lines on the rule to the left of the Vernier zero in whole inches plus fractions of $^1/_{16}$ths of an inch to the line on the Vernier that matches a line on the rule in $^1/_{128}$ths of an inch to find the total length.

5. To read the Vernier caliper in metric units, use the lower scale. The rule is divided into centimeters and millimeters. Each division on the Vernier is equal to $^9/_{10}$ mm, and the difference between one division on the rule and one division on the Vernier is $^1/_{10}$ mm. To derive total length, take the reading on the rule to the left of the Vernier zero for whole number digits in mm, and take the line on the Vernier opposite to a line on the rule for $^1/_{10}$ mm. The reading may also be reported in centimeters to the hundredths place.

6. To calculate percent sag:

$$\% \text{ sag } = \frac{\text{height in container} - \text{height out of container}}{\text{height in container}} \times 100$$

■ VISCO/AMYLO/GRAPH*

Measures

Apparent viscosity as a function of time and temperature for stirred starch dispersions.

Uses

Starch paste behavior. This instrument is primarily used to study gelatinization, breakdown, and setback of starch pastes.

Procedure

1. Set the pen (A) on zero Brabender units (see Figure 20.19). Advance the chart paper until the pen is on the zero time line.

2. Pour test slurry into the bowl (B). Insert bowl into heating unit (C) and rotate bowl by hand to locate key. Replace stirrer.

3. Swing head forward and lower into position carefully with knob (D). Position stirrer and lock into place with coupling pins (E) as head is lowered.

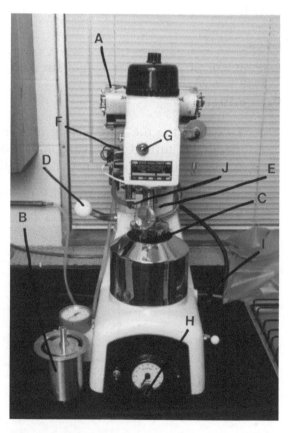

Figure 20.19   Visco/amylo/GRAPH.

*   C.W. Brabender Instruments, Inc.

Heat Cycle:

4. Preset time. The temperature rises at 1.5°C/min, so most starches require 45 to 50 min.

5. Turn on the main switch (small white-topped lever on the right side of the instrument).

6. Set the thermoregulator gear shift lever (F) to neutral.

7. Turn on the thermoregulator light switch.

8. Adjust the thermoregulator (G) to the temperature of the sample.

9. Set the thermoregulator gear shift lever to "UP" for programmed heating of 1.5°C/min. Cooling regulator should be in the middle position.

10. Push the red button on the timer (H) to begin the heating cycle and turn on the alarm switch.

11. Rotate the speed knob (I) on the right of the instrument to obtain correct r/min (75 r/min = standard).

Plateau:

12. Reset time for 15 min.

13. Place thermoregulator switch (F) to the "zero" position.

14. Push the red button (H) and turn on the switch.

Cooling Cycle:

15. Turn on the water supply.

16. Move the thermoregulator gear shift lever (F) to the "DOWN" position.

17. Place cooling regulator in "CONTROLLED" position.

18. Immerse cooling probe (J).

19. Reset time.

20. Depress timer button (H) to start cooling cycle at 1.5°C/min.

The resulting curve is a plot of temperature/time vs. apparent viscosity in Brabender units and is referred to as a Brabender amylogram.

■ WATER ACTIVITY SYSTEM*

Measures

Water activity based on dew point. A thermocouple detects the condensation temperature on a cooled mirror, which is related to the moisture.

---

* CX-2; Decagon Devices, Inc.

Uses

Indication of the free water in a food:

$$a_w = p_{food}/p_{water} = \% \text{ ERH}/100$$

where p_{food} = water vapor pressure of water over the food, p_{water} = water vapor pressure over pure water, and ERH = equilibrium relative humidity.

Procedures

1. Turn on power switch (see Figure 20.20). The instrument needs to warm up for 15 to 60 min.

2. Prepare samples in plastic sample dishes. Do not fill sample cup more than half full. Lids can be used to prevent moisture loss during storage.

3. Pull out sample drawer and insert sample cup.

4. Close drawer and turn knob from "OPEN/LOAD" to "READ."

5. Take readings of the a_w and temperature of the sample after consecutive readings are less than 0.001 apart, which indicates that equilibrium has been achieved (<5 min). The CX-2 will beep, and decimal points on display will blink after each measurement process.

6. Remove sample. Never leave a sample in the CX-2.

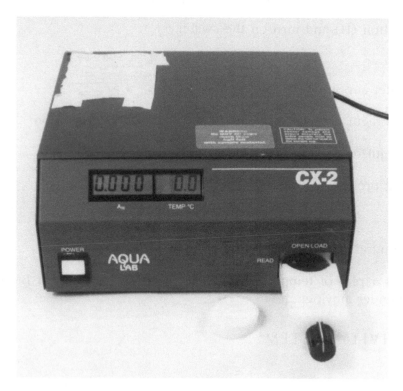

Figure 20.20 Water activity system.

APPENDIX

■ DIRECTIONS FOR SETTING UP DIFFERENCE TESTING EXPERIMENT

1. For the experiment on page 38, put 15 ml of each test solution into coded 2-oz sample cups.

2. Coding should be done by the instructor or by a laboratory assistant rather than by students who will taste the samples.

3. A 5% solution of citric acid may be used to acidify the samples of juice.

4. For 16 students, 250 ml of each test juice should be provided. Quantities of each test juice by three-digit code and by percent of 5% citric acid solution in juice (by volume) are as follows:

Code	% of 5% Citric Acid Solution in Apple Juice	Test Juice (ml)
545	0	250
390	1	250
923	1	250
517	0	250
886	0	250
792	0	250
459	1	250
609	2	250
904	3	250
534	4	250
269	1	250
919	3	250
109	2	250
512	2	250
204	0	250
843	8	250
Reference	2	250

Total Juice for Sixteen Students

Citric Acid Solution in the Test Juice (%)	Volume of Citric Acid Solution (ml)	Volume of Test Juice (ml)
0	0	1250
1	10	1000
2	20	1000
3	15	500
4	10	250
8	20	250
Totals	75	4250

INDEX

T - #0711 - 101024 - C0 - 280/208/8 - PB - 9780849312939 - Gloss Lamination